T0203131

THE
REAL PROJECTIVE
PLANE

THIRD EDITION

By the same author

NON-EUCLIDEAN GEOMETRY

UNIVERSITY OF TORONTO PRESS

REGULAR POLYTOPES

DOVER, NEW YORK

INTRODUCTION TO GEOMETRY

WILEY, NEW YORK

PROJECTIVE GEOMETRY

SPRINGER-VERLAG, NEW YORK

TWELVE GEOMETRIC ESSAYS

SOUTHERN ILLINOIS UNIVERSITY PRESS

REGULAR COMPLEX POLYTOPES

CAMBRIDGE UNIVERSITY PRESS

H.S.M. COXETER

THE
REAL PROJECTIVE
PLANE

WITH AN APPENDIX FOR MATHEMATICA® BY GEORGE BECK

MACINTOSH VERSION

(Diskette provided)

THIRD EDITION

Springer-Verlag
New York Berlin Heidelberg London Paris
Tokyo Hong Kong Barcelona Budapest

H.S.M. Coxeter
Department of Mathematics
University of Toronto
Toronto, Ontario M5S 1A1
CANADA

George Beck
117 Lyndhurst Avenue
Toronto, Ontario M5R 2Z8
CANADA

Library of Congress Cataloging-in-Publication Data
Coxeter, H.S.M. (Harold Scott Macdonald), 1907–
    The real projective plane / H.S.M. Coxeter; with an appendix for
Mathematica by George Beck.—3rd ed.
        p.   cm.
    Includes bibliographical references (p.   ) and index.
    ISBN 978-1-4612-7647-0   ISBN 978-1-4612-2734-2 (eBook)
    DOI 10.1007/978-1-4612-2734-2

    1. Geometry, Projective. 2. Geometry, Projective—Data
processing.   I. Title.
QA471.C68   1992
516′.5—dc20   92-22637

Printed on acid-free paper.

First edition published in 1949 by the McGraw-Hill Book Company, Inc. Second edition
published in 1955 by Cambridge University Press.

Production managed by Dimitry L. Loseff; manufacturing supervised by Vincent Scelta.
Typeset by Asco Trade Typesetting Ltd., Hong Kong.

9 8 7 6 5 4 3 2 1

Preface to the Third Edition

Along with many small improvements, this revised edition contains van Yzeren's new proof of Pascal's theorem (§1.7) and, in Chapter 2, an improved treatment of order and sense. The Sylvester-Gallai theorem, instead of being introduced as a curiosity, is now used as an essential step in the theory of harmonic separation (§3.34). This makes the logical development self-contained: the footnotes involving the References (pp. 214–216) are for comparison with earlier treatments, and to give credit where it is due, not to fill gaps in the argument.

H.S.M.C.

November 1992

Preface to the Second Edition

Why should one study the *real* plane? To this question, put by those who advocate the complex plane, or geometry over a general field, I would reply that the real plane is an easy first step. Most of the properties are closely analogous, and the real field has the advantage of intuitive accessibility. Moreover, real geometry is exactly what is needed for the projective approach to non-Euclidean geometry. Instead of introducing the affine and Euclidean metrics as in Chapters 8 and 9, we could just as well take the locus of 'points at infinity' to be a conic, or replace the absolute involution by an absolute polarity.

Apart from the correction of many small errors, the changes made in this revised edition are chiefly as follows: von Staudt's proof that $AA'BB' \barwedge A'AB'B$ (2·71) has been adapted to yield the quadrangular involution (4·71). The first axiom of order has been weakened (3·11). More satisfactory proofs have been given for Hesse's theorem (5.55), for von Staudt's converse of Chasles's theorem (5·71), for Archimedes' axiom (10·22), and for Enriques's fixed-point theorem (10·62). There is also an improved treatment of degenerate polarities (5·9), of the inside and outside of a conic (6·32), of Desargues's involution (6·72), of the nine-point conic (6·81), of the condition for a quadrangle to be convex with respect to a line (7·55), and of Klein's classification of geometries according to the groups of transformations under which their properties are invariant (8·10).

I wish to express my gratitude to many readers of the first edition who sent useful suggestions, to W.O.J. Moser for helping with the proofs, and to the Syndics of the Cambridge University Press for undertaking the new edition.

H.S.M.C.

November 1954

Preface to the First Edition

This introduction to projective geometry can be understood by anyone familiar with high-school geometry and algebra. The restriction to real geometry of two dimensions makes it possible for every theorem to be illustrated by a diagram. The early books of Euclid were concerned with constructions by means of ruler and compasses; this is even simpler, being the geometry of the ruler alone. The subject is used, as metrical geometry was by Euclid, to reveal the development of a logical system from primitive concepts and axioms. Accordingly, the treatment is mainly synthetic; analytic geometry is confined to the last two of the twelve chapters.

The strict axiomatic treatment is followed far enough to show the reader how it is done, but is then relaxed to avoid becoming tedious. Continuity is introduced in Chapter 3 by means of an unusual but intuitively acceptable axiom. A more thorough treatment is reserved for Chapter 10, at which stage the reader may be expected to have acquired the necessary maturity for appreciating the subtleties involved.

The spirit of the book owes much to the great *Projective Geometry* of Veblen and Young. That dealt with geometries of various kinds in any number of dimensions; but the present book may be found easier because one particular geometry has been extracted for detailed consideration. Chapters 5 and 6 constitute what is perhaps the first systematic account in English of von Staudt's synthetic approach to polarities and conics as amplified by Enriques: A polarity is defined as an involutory point-to-line correspondence preserving incidence, and a conic as the locus of points that lie on their polars, or the envelope of lines that pass through their poles. This definition for a conic gives the whole figure at once and makes it immediately self-dual, a locus and an envelope, whereas Steiner's definition assigns a special role to two points on the conic, obscuring its essential symmetry. Moreover, the restriction to real geometry makes it desirable to consider not only the hyperbolic polarities which determine conics but also the elliptic polarities which do not. The latter are important because of their applica-

tion to elliptic geometry. (In complex geometry this distinction is unnecessary, for an elliptic polarity determines an imaginary conic.) The
linear construction for the polar of a given point (5·64) was adapted
from a question in the Cambridge Mathematical Tripos, 1934, Part II,
Schedule A.

The treatment of conics is followed in Chapter 8 by a description of
affine geometry, where one line of the projective plane is singled out as
a *line at infinity*, enabling us to define parallel lines. It is interesting to
see how much of the familiar content of metrical geometry depends
only on incidence and parallelism and not on perpendicularity. This
includes the theory of area; the distinction between the ellipse, parabola and hyperbola; and the theory of diameters, asymptotes, etc. The
further specialization to Euclidean geometry is made in Chapter 9 by
singling out an *absolute involution* on the line at infinity.

Chapter 10 introduces a revised axiom of continuity for the projective line, so simple that only eight words are needed for its enunciation. (This has not been published elsewhere save as an abstract in the
Bulletin of the American Mathematical Society.) Chapter 11 develops
the formal addition and multiplication of points on a conic and the
synthetic derivation of coordinates. Finally, Chapter 12 contains a
verification that the plane of real homogeneous coordinates has all the
properties of our synethetic geometry. This proves that the chosen
axioms are as consistent as the axioms of arithmetic.

Almost every section of the book ends with a group of problems
involving the latest ideas that have been presented. All the difficult
problems are followed by hints for solving them. The teacher can ren-
der them more difficult by taking them out of their context or by omitting the hints.

I take this opportunity for expressing my thanks to H.G. Forder and
Alan Robson for reading the manuscript and suggesting improvements; also to Leopold Infeld and Alex Rosenberg for helping with the
proofs.

H.S.M.C.

February 1949

Contents

Chapter 6. CONICS

Chapter 7. PROJECTIVITIES ON A CONIC

Chapter 8. AFFINE GEOMETRY

Chapter 9. EUCLIDEAN GEOMETRY

Chapter 10. CONTINUITY

Chapter 11. THE INTRODUCTION OF COORDINATES

Chapter 12. THE USE OF COORDINATES

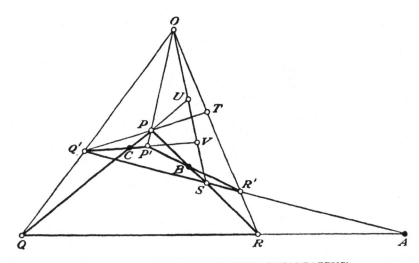

DESARGUES' THEOREM DEDUCED FROM PAPPUS'

CHAPTER 1

A Comparison of Various Kinds of Geometry

1·1 Introduction. The ordinary geometry taught in school, deal-ing with circles, angles, parallel lines, similar triangles and so on, is called *Euclidean geometry* because it was first collected into a system-atic account by the Greek geometer Euclid, who lived about 300 B.C. His treatise, *The Elements*, is one of the most famous books in the world; probably the Bible is its only rival in the number of copies made and the number of languages into which it has been translated. With a few unimportant changes it is still suitable for the instruction of the young.

During the nineteenth century there was a tendency to extract from Euclidean geometry certain ideas of a particularly simple nature, espe-cially ideas that did not involve measurement of distance or angle, and to use these for building up more general systems, notably *affine* geom-etry and *projective* geometry. The meaning of these terms will be clear after we have examined certain kinds of projection. For that purpose we shall need some intuitive notions of solid geometry; but after the present chapter we shall be concerned solely with plane geometry.

These new systems are said to be more general because, besides throwing fresh light on Euclidean geometry itself, they are capable of extension in other directions by the introduction of new kinds of measurement. Affine geometry can be developed into Minkowski's geometry of the space-time continuum considered in the special theory of relativity, and projective geometry can be developed into the various kinds of non-Euclidean geometry that are relevant to more modern ideas of relativistic cosmology. This remark is intended merely to show why it is worthwhile to study these fundamental geometries; the extensions themselves are beyond the scope of this book.

1·2 Parallel projection. Two figures in distinct planes are said to be derived from each other by *parallel projection* if corresponding points can be joined by parallel lines.* (This is essentially what happens when the sun casts a shadow on the ground; e.g. when a circular coin casts an elliptic shadow, the lines joining each point of the circle to its shadow on the ellipse are parallel.) If the two planes are parallel, the two figures will be exactly alike (congruent); otherwise they may have somewhat different shapes, but straight lines remain straight, tangents to curves remain tangent, parallel lines remain parallel, bisected segments remain bisected, equal areas remain equal. In other words, the properties of straightness, tangency, parallelism, bisection, and equality of area are *invariant* under parallel projection. Such properties are the subject-matter of *affine* geometry. (The use of the word *affine* is due to the Swiss mathematician Euler, 1707–83.)

On the other hand, the content of *projective* geometry is still more restricted, being confined to those properties (such as straightness and tangency) which remain invariant under central projection.

1·3   Central projection.† Two figures in distinct planes are said to be derived from each other by *central projection* if corresponding points can be joined by *concurrent* lines, all passing through a fixed point L. (This is essentially what happens when a lamp casts a shadow on a wall or on the floor. The circular rim of a lampshade usually gives a larger circular or elliptic shadow on the floor and a hyperbolic shadow on the nearest wall.) If the two planes are parallel, the two figures will be similar and the invariant geometry will again be affine. So we shall assume the two planes to be non-parallel; then the plane through L parallel to one of the two planes will meet the other in a definite line called the *vanishing line* for a reason that will soon be explained.

Figure 1·3A represents a box standing on a table with a lamp suspended inside the lid at L. A figure is drawn opaquely on the transparent vertical side oP of the box so as to cast a shadow on the horizontal plane of the table top. Clearly, the shadow is derived from the original figure by central projection from L. In general, two intersecting lines project into two intersecting lines, just as in the case of parallel projection; but an exception arises when the given lines intersect on the special line o, which lies in the horizontal plane through L. Such lines, say AP and AQ, project into *parallel* lines p and q through P and Q, both parallel to LA. Conversely, any two parallel lines on the table top are each coplanar with the parallel line through L, say LA; therefore, unless they are parallel to o, they must be projected images of two lines through a definite point A on o.

---

* We shall always use the word *line* in the sense of a straight line of unlimited extent.

† Cremona 1960, p. 3. (All references are listed in the bibliography on pp. 214–216.)

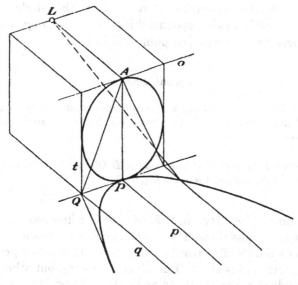

Fig. 1·3A

This process of central projection relates points in the vertical and horizontal planes in such a way that the join of two corresponding points always passes through L. Points in the vertical plane below the table top project into points inside the box, and points X above o project into points behind the box (on XL produced), although now the notion of a shadow breaks down. Thus the only points that have no images are those on the line o, which is consequently called the *vanishing line*.

A circle in the vertical plane is joined to L by a cone (usually an oblique circular cone); thus it projects into a conic section, or *conic*. If the circle does not meet o, the conic is a closed oval curve, viz. an *ellipse*. If the circle cuts o in two distinct points, the conic is a *hyperbola*, which has two branches arising from the arcs below and above o. (The latter branch is behind the box.) Finally, if the circle touches o (as in Fig. 1·3A), the conic is a *parabola*, which has only one branch but is not closed. Note that other tangents to the circle (such as t) project into tangents to the conic. We shall not be surprised to find that the role of conics in projective geometry is almost as vital as the role of circles in Euclidean geometry, though actually we shall not make use of them till Chapter 6.

We have taken the plane oP to be vertical for simplicity. An oblique plane could be used just as well.

These ideas formed the foundation for the work of a remarkable Frenchman, Jean Victor Poncelet (1788–1867), who fought in Napoleon's Russian campaign (1812) until the Russians took him prisoner. Being deprived of all books, he decided to reconstruct the whole science of geometry. The result was his epochmaking *Traité des propriétés projectives des figures*,* which was first published in 1822.

EXERCISES

1. Lines parallel to the vanishing line remain parallel after projection, but any other parallel lines project into intersecting lines. Show how to determine their point of intersection.

2. Let a circle cut the vanishing line in A and B. Observe how the tangents at A and B project into the asymptotes of the hyperbola.†

1·4   The line at infinity. In §1·2 we defined affine geometry as consisting of those propositions of Euclidean geometry which retain their meaning and validity after parallel projection; thus every proposition of affine geometry holds also in Euclidean geometry, but other propositions of Euclidean geometry (such as Euclid I, 1 and 5) are essentially meaningless in affine geometry. Somewhat similarly, projective geometry includes all propositions of affine geometry that retain their meaning and validity after central projection; but this is not the whole story. Some statements are true in projective geometry but false in affine geometry. The most important instance is: 'Any two lines in a plane have a point of intersection.' This fails in affine geometry because the two lines might be parallel. The projective statement is validated by inventing a new kind of point so as to be able to say that parallel lines have a common *point at infinity*: the projected image of a point on the vanishing line. This vitally important concept is due to the great German astronomer Kepler (1571–1630).

We often think of a line as consisting of all the points on it, i.e. a *range* of points. It is equally useful to think of a point as consisting of all the lines through it, i.e. a *pencil* of lines. Statements about points are easily translated into statements about pencils; e.g. 'Two points lie on just one line' becomes 'Two pencils contain just one common line'. Lines through A (in the plane oP of Fig. 1·3A) project into parallel lines (such as p and q) on the horizontal plane. If we agree to call these a *pencil of parallels*, we may say that a pencil always projects into a pencil. When statements about such pencils are translated back into statements about points, we have to admit points at infinity as well as ordinary points. In fact, we call the pencil of parallels a point at infinity, denote it by A', and call it the projected image of the ordinary point A on the vanishing line o.

---

* Poncelet 1865.

† Readers who have not studied the hyperbola should omit this exercise.

In this manner we have extended the meaning of the word *point* so as to be able to say that any two coplanar lines intersect in a point. Similarly, we extend the meaning of the word *line* so as to be able to say that any two planes intersect in a line. If the two planes happen to be parallel, this is a *line at infinity*. Since we have agreed to call A' the projected image of A, all points at infinity in the plane *pq* lie also in the parallel plane *Lo* and form a range on the line at infinity, which is the intersection of the two parallel planes, i.e. the projected image of the vanishing line *o*. However, there seems to be a paradox here: we have agreed that the point at infinity A' is really only another name for the pencil of lines parallel to *p*, and yet we have declared that it lies in the plane *Lo*, which certainly does not contain these lines. The explanation is that for brevity we have oversimplified the account. For a complete treatment of 'ideal elements' we should have to consider the whole space, using *bundles* of lines and planes (i.e. all the lines and planes through a given point or parallel to a given line*) instead of pencils of lines. Then a 'point' is said to lie in a plane if the plane belongs to the bundle; and in the case of a bundle of parallels this merely means that the plane contains a line in the direction of the bundle. When we restrict consideration to a single plane, the bundle is replaced by a pencil, all the lines of an ordinary pencil contain different points at infinity (which belong equally to the respectively parallel lines of any other ordinary pencil), and all these points at infinity are to be regarded as a range on the line at infinity. We can treat the points at infinity just like any other range of points so long as we are dealing with properties that are invariant under central projection.

By introducing these new elements we have enlarged the affine plane (in which both the affine and Euclidean geometries operate) so as to obtain the *projective plane*, which has simpler properties of incidence. For other purposes we might choose different ways to enlarge the affine plane, but the corresponding geometries would be outside the scope of this book.

EXERCISE

Which of the following propositions belong to Euclidean geometry, which to affine, and which to projective?
- (*a*) Four lines of general position have six points of intersection.
- (*b*) If lines *AB* and *CD* intersect, then *AC* and *BD* intersect.
- (*c*) The diagonals of a parallelogram bisect each other.
- (*d*) The three medians of a triangle have a common point.
- (*e*) The three altitudes of a triangle have a common point.
- (*f*) The angle in a semicircle is a right angle.

---

* Any two coplanar lines (intersecting or parallel) determine such a bundle, which contains the intersections of all the planes through one line with all the planes through the other (see Coxeter 1965, p. 166). The analytic aspect of ideal elements is neatly described by Hardy (1925, Appendix III).

1·5 Perspective triangles. If the three sides of one triangle are parallel to the three sides of another, the two triangles are, of course, similar. Thus the part of Euclid's Book VI that deals with similar *and similarly situated* figures belongs to affine geometry. The following theorem is easily proved in this manner:

1·51 *Let PQR and P'Q'R' be two triangles (in the affine plane) with QR parallel to Q'R' and RP parallel to R'P', while the joins PP', QQ', RR' are concurrent. Then PQ is parallel to P'Q'.*

Proof: Let O be the common point of PP', QQ', RR', as in Fig. 1·5A. Applying Euclid VI.2 to triangles QOR and Q'OR', ROP and R'OP', we have

$$\frac{OQ'}{OQ} = \frac{OR'}{OR} = \frac{OP'}{OP}.$$

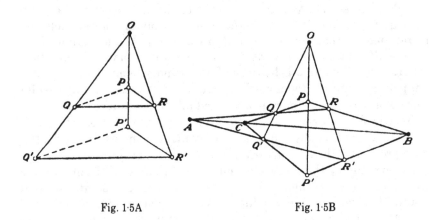

Fig. 1·5A   Fig. 1·5B

Hence the triangles POQ and P'OQ' are similar, and their corresponding sides PQ and P'Q' are parallel.

This affine theorem has a projective generalization that is very important:

1·52 Desargues's two-triangle theorem. *If two triangles have corresponding vertices joined by concurrent lines, then the intersections of corresponding sides are collinear.*

In other words, if PP', QQ', RR' all pass through one point O, as in Fig. 1·5B, then the intersections

$$A = QR \cdot Q'R', \quad B = RP \cdot R'P', \quad C = PQ \cdot P'Q'$$

all lie on one line.

Proof: Imagine the figure drawn on the plane oP of Fig. 1·3A, with AB for the vanishing line. By projection onto the horizontal plane we obtain two triangles having two pairs of parallel sides, as in 1·51. We conclude that their remaining sides are parallel. Since the sides PQ and $P'Q'$ of the original triangles project into these parallel lines, their point of intersection C must lie on the vanishing line AB, as required.

Of course, 1.51 is just a special case of 1·52, obtained by taking AB to be the line at infinity. In projective geometry the line at infinity is treated like any other line; therefore, if the corresponding sides meet in collinear points in this special case, they must still do so for *any* position of AB. In this spirit, instead of projecting the figure onto another plane, we could say, 'Make the plane affine by choosing AB as the line at infinity.'

Since we shall eventually take Desargues's theorem (our 1·52) as an axiom, it seems worthwhile to given an alternative proof: von Staudt's projective three-dimensional proof. First, we observe that the theorem is almost obvious when applied to two triangles in distinct planes; for in that case the points A, B, C all lie in the plane PQR and also in the plane $P'Q'R'$, and therefore they all lie on the line of intersection $PQR \cdot P'Q'R'$. The theorem for triangles in one plane arises as a limiting case; but if we prefer not to use such considerations of continuity, we may proceed as follows: Take any two points S and S' on a line through O outside the plane of the two given triangles, so that the four lines PP', QQ', RR', SS' all pass through O. Since P, P', S, S' all lie in one plane OPS, the lines PS, $P'S'$ meet in a point P_1 (possibly at infinity); similarly, QS meets $Q'S'$ in a point Q_1, and RS meets $R'S'$ in a point R_1. Applying the 'obvious' version of the theorem to the triangles QRS, $Q'R'S'$, which lie in distinct planes, we see that the points of intersection

$$R_1 = RS \cdot R'S', \quad Q_1 = SQ \cdot S'Q', \quad A = QR \cdot Q'R'$$

are collinear. Thus A lies on Q_1R_1; similarly, B on R_1P_1 and C on P_1Q_1. Hence the three points A, B, C, lying in the plane $P_1Q_1R_1$ as well as in the plane PQR, must lie on the line of intersection $PQR \cdot P_1Q_1R_1$.

The reader will notice that both the above proofs involve constructions outside the plane. This use of the third dimension is inevitable, unless we take Desargues's theorem (or some equivalent statement) as an axiom.

Girard Desargues (1593–1662) was an architect of Lyons. He discovered not only the above theorem but also several others, which we shall use later, especially in connexion with conics. His treatise of 1639 was not well received during his lifetime, partly because of his obscure style; he introduced about seventy new terms, of which only *involution* has survived.

EXERCISE

By taking the corresponding vertices R and R' to be points at infinity, deduce from 1·52 the converse of 1·51: *If two triangles have parallel sides, the joins of their corresponding vertices are concurrent or parallel.*

1·6   The directed angle, or cross. One of the concepts to which we shall be led in Chapter 9 is that of *angle*: not the customary 'angle between two rays' but an 'angle between two *lines*', which is subtly different. From this standpoint, $\langle AOB \rangle$ means the angle through which a variable line has to be turned, in the counterclockwise sense, in order to pass from the position AO to the position OB. Thus $\langle AOB \rangle$ is not altered by shifting A along AO, or B along OB, even beyond O. Such angles may be measured in degrees or radians, but they cannot be negative and are always less than 180° or π. The angles $\langle AOB \rangle$ and $\langle BOA \rangle$ are supplementary, as we see in Fig. 1·6A. Here are some of Euclid's propositions expressed in terms of these directed angles:

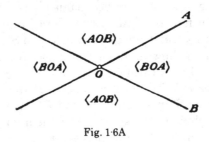

Fig. 1·6A

I.5. *If triangle ABC has equal sides AB and AC, then*

$$\langle ABC \rangle = \langle BCA \rangle.$$

I.13. $\langle AOB \rangle + \langle BOA \rangle = \pi$.
I.27. *If* $\langle PAB \rangle = \langle QBA \rangle$, *then AP is parallel to BQ.*
III.21 and 22. *If A, B, C, D lie on a circle, then*

$$\langle ACB \rangle = \langle ADB \rangle.$$

III.32. *If AT is the tangent at A to the circle ABC, then*

$$\langle ACB \rangle = \langle TAB \rangle.$$

This kind of angle was invented independently by R.A. Johnson (1929, pp. 12–15) and D.K. Picken (1925, pp. 45–55; see also Forder 1931, pp. 7–20; 1947, pp. 227–233).

EXERCISES

1. Show that $\langle ACD \rangle = \langle BCD \rangle$ if A, B, C are collinear.

2. Show that $\langle AOB \rangle = \langle BOA \rangle$ if OA and OB are perpendicular.

3. Show that $\langle AOX \rangle = \langle XOB \rangle$ if OX is the internal or external bisector of the angle $\langle AOB \rangle$.

1·7 Hexagramma mysticum. The religious philosopher Blaise Pascal (1623–1662) discovered his theorem of the *hexagramma mysticum* when he was sixteen years old. It was published in 1640 with the help of Desargues, under the title *Essay pour les Coniques* (Pascal 1908, p. 252). This single page is all that remains of his extensive treatise, which was praised by G.W. Leibniz but was then lost. Of the many attempts that have been made to reconstruct Pascal's original proof, the most elegant is the following, which has been kindly provided by J. van Yzeren of Eindhoven, the Netherlands.

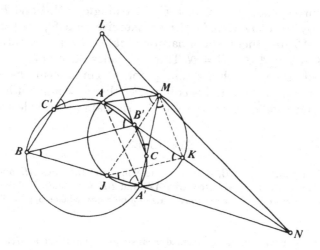

Fig. 1·7A

1·71 Pascal's theorem. *If a hexagon is inscribed in a conic, the pairs of opposite sides intersect in three collinear points.*

Proof: Since any conic can be projected into a circle (and *vice versa*), no generality will be lost by our taking the conic to be a circle, as in Fig. 1·7A. Here the hexagon is $AB'CA'BC'$, so that the points to be proved collinear are

$$L = BC' \cdot CB', \quad M = CA' \cdot AC', \quad N = AB' \cdot BA'.$$

(The hexagon is not necessarily convex!)

We first dispose of the case when the hexagon has more than one pair of parallel sides. If AB' is parallel to BA', and BC' to CB', then the circle has equal directed arcs

$$AA' = BB' = CC',$$

whence AC' is parallel to CA'; the points L, M, N all lie on the line at infinity.

We may now assume that at least two of the points L, M, N are ordinary (not at infinity), and no generality will be lost by taking them to be L and M. Let J and K be the second intersections of the lines BA' and AB' with the circle $AA'M$.[*]

Since $\langle AKJ \rangle = \langle AA'J \rangle = \langle AA'B \rangle = \langle AB'B \rangle$,

$\qquad\qquad\qquad\qquad$ JK is parallel to BB'.

Since $\langle AMJ \rangle = \langle AA'J \rangle = \langle AA'B \rangle = \langle AC'B \rangle$,

$\qquad\qquad\qquad\qquad$ JM is parallel to $BC' = BL$

Since $\langle A'MK \rangle = \langle A'AK \rangle = \langle A'AB' \rangle = \langle A'CB' \rangle$,

$\qquad\qquad\qquad\qquad$ KM is parallel to $B'C = B'L$.

Thus corresponding sides of the two triangles JKM and $BB'L$ are parallel. By the converse of 1.51 (see Exercise on p. 8), the three lines JB, KB', ML are concurrent or parallel; that is, ML passes through the point $JB \cdot KB' = A'B \cdot AB' = N$. This concludes the proof.

Had Leibniz seen such a simple proof, he could easily have reconstructed it and shown it to his colleagues; then it would not have been lost. We may safely infer that Pascal's original proof was less elegant.

EXERCISES

1. For four points A, A', B, B' on a circle, show that the lines AB' and BA' are parallel if and only if the directed arcs AA' and BB' are equal. Deduce that any three equal chords of a circle are diagonals of a hexagon whose 'opposite' sides are parallel.

2. How many hexagons can be formed from six given points on a circle?

1·8 An outline of subsequent work. In the next chapter we shall make a fresh start, considering real projective geometry as a self-contained system, defined by its own peculiar axioms. We have the satisfaction of knowing that it is *consistent* (i.e. that its axioms cannot lead to any contradictory statements) because we can obtain a 'model' of it

---

[*] The same auxiliary circle was used, with different auxiliary points, by Guggenheimer 1967, pp. 104–105.

by adding the line at infinity to the affine plane. Here we are taking for granted the consistency of Euclidean geometry, which includes affine geometry.

This investigation of projective geometry will be continued throughout Chapters 2–7. Then we shall derive affine geometry in Chapter 8 and Euclidean in Chapter 9. At that stage we shall have returned to our starting point, ready to deal in a more sophisticated manner with the difficult subject of continuity (Chapter 10). Finally, in Chapters 11 and 12 we shall see how these *synthetic* geometries lead to , and can be derived from, *analytic* geometry.

Chapter 2

Incidence

The geometry considered in this book is called *real*, because if we chose to work it out analytically, the coordinates would be *real* numbers, whereas otherwise they might have been complex numbers, or the 'numbers' of a finite arithmetic (Galois field),* or something still more bizarre. However, the present chapter deals with those properties of the projective plane which depend only on the simple processes of joining and intersection and which are consequently valid in the other geometries mentioned above, as well as in real geometry. These properties include the principle of duality, perspectivity, and harmonic conjugacy. Many of the ideas can be traced back to Desargues (who defined harmonic conjugates by dividing a segment internally and externally in the same ratio), but their essentially projective nature was first understood by an extraordinarily talented German, von Staudt (1798–1867).

2·1 Primitive concepts. In a logical development of geometry, each definition of an entity or relation involves other entities or relations; therefore some entities and relations, the *primitive concepts*, must remain undefined. Similarly, the proof of each proposition uses other propositions; therefore some propositions, the *axioms*, must remain unproved. In practice, the primitive concepts should have some intuitive significance, some interpretation in which the axioms are seen to be true; otherwise, we should be playing a meaningless game.

A basis for the system of real projective geometry may be chosen in many different ways. It seems simplest to take as primitive concepts *point*, *line*, *incidence*, and *separation*.

There is no harm in picturing a point as the idealized limit of smaller and smaller material dots ('position without magnitude') and a line as the idealized limit of a material line drawn with a sharp pencil on smooth paper against a straight ruler. Of course, a microscope would

---

* Robinson 1940, pp. 87–89, 106–108.

reveal imperfections in any material line, but the geometrical object is supposed to be perfectly thin and straight and infinitely extended.

A point and a line may or may not be *incident*. When they are, we say that the point lies on the line or that the line passes through the point. A line passing through two points is called their *join*, and a point lying on two lines is called their *intersection*.

The relation of *separation* applies to two pairs of points on a line or to two pairs of lines through a point. If four such points (or lines) A, B, C, D occur in that order, we say that A and C *separate* B and D and write for brevity $AC//BD$.

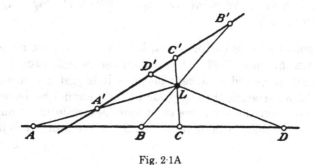

Fig. 2·1A

This relation has a less straightforward meaning in complex geometry and no meaning whatsoever in finite geometries, but it does properly belong to *real* projective geometry, for it is invariant under central projection (see Fig. 2·1A, where $AC//BD$ and $A'C'//B'D'$). On the other hand, the simpler notion of B lying 'between' A and C, which belongs to affine geometry, is not invariant. (Visibly, B' is not between A' and C'.) In fact, the kind of order that belongs to real projective geometry is not serial but cyclic. We shall return to these considerations in the next chapter.

Although the above description of the primitive concepts helps our imagination and thus suggests what axioms are appropriate, we must take care never to use any of these intuitive ideas in our proofs. The only properties to be assumed are those actually stated in the axioms.

The two processes of joining and intersection, which emerge from the relation of incidence, somewhat resemble the processes of addition and multiplication in algebra and are sometimes denoted by the same symbolism. We shall adopt the multiplicative symbol $a \cdot b$ for the intersection of lines a and b, but the join of points A and B will be denoted by the familiar symbol AB, rather than the more startling $A + B$. These symbols are easy to combine; e.g. the intersection of AB and CD is $AB \cdot CD$, while the join of $a \cdot b$ and $c \cdot d$ is $(a \cdot b)(c \cdot d)$.

2·2 The axioms of incidence.

2·21 *There exist a point and a line that are not incident.*

2·22 *Every line is incident with at least three distinct points.*

2·23 *Any two distinct points are incident with just one line.*

2·24 *Any two lines are incident with at least one point.* (It follows that any two distinct lines are incident with *just* one point.)

2·25 *If three lines PP', QQ', RR' are all distinct and incident with one point, then the three points $QR \cdot Q'R'$, $RP \cdot R'P'$, $PQ \cdot P'Q'$ are all incident with one line.*

Such a set of axioms (using point, line, and incidence as primitive concepts) was given in 1899.* They are all very simple except the fifth, which we are prepared to accept because it is just a restatement of Desargues's two-triangle theorem (our 1·52). (It cannot be deduced from the four simple axioms; for there exist 'non-Desarguesian' geometries† that satisfy 2·21--2·24 without satisfying 2·25). We proceed to prove its converse:

2·26 *If two triangles have corresponding sides intersecting in collinear points, then the joins of corresponding vertices are concurrent.*
Proof: Using the same notation as in 1·52, we have two triangles PQR and $P'Q'R'$ whose corresponding sides intersect in the three collinear points A, B, C, and we wish to prove that the line RR' passes through the point $O = PP' \cdot QQ'$ (see Fig. 2·2A). This is an immediate consequence of Desargues's two-triangle theorem itself, as applied to the triangles AQQ' and BPP', whose joins of corresponding vertices all pass through C, while their intersections of corresponding sides are O, R', R.

EXERCISES

1. Give detailed proofs of the following simple theorems, pointing out which axioms are used:
(a) Every point is incident with at least three lines.
(b) Any two distinct lines are incident with just one point.

---

* Pier 1899, pp. 6–22, Postulates I–XIII.
† See, for example, Robinson 1940, pp. 126–128. Such geometries can occur only in two dimensions. Pieri avoided the assumption of 2·25 by working in three dimensions.

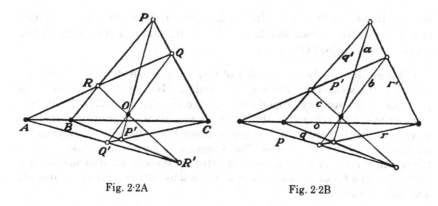

Fig. 2·2A   Fig. 2·2B

2. The two triangles PQR and $P'Q'R'$ of Fig. 2·2A are conveniently said to be *perspective* from the *centre O* and *axis ABC*. If three triangles are all perspective from the same centre, prove that the three axes are concurrent. (*Hint:* Let the three axes be A_1B_1, A_2B_2, A_3B_3. Apply 2·26 to triangles $A_1A_2A_3$ and $B_1B_2B_3$.)

3. Show that the 10 points and 10 lines of the Desargues configuration (Fig. 2·2A or B) may be renamed $P_{12}, \ldots, P_{45}$ and $p_{12}, \ldots, p_{45}$ in such a way that P_{ij} and p_{kl} are incident whenever the numbers represented by i, j, k, l are all different. (*Hint:* Let $PQRP'Q'R'$ be called $P_{14}P_{24}P_{34}P_{15}P_{25}P_{35}$. This notation arises from the figure of 5 points in space with the 10 lines and 10 planes that join them. Taking the section by a plane of general position, we obtain P_{12} and p_{12} as sections of the line 12 and plane 345. The five points 1, 2, 3, 4, 5 may be identified with the P_1, Q_1, R_1, S, S' of p. 7.)

4. Show that the same 10 points and 10 lines may be regarded (in six ways) as consisting of two pentagons so situated that consecutive sides of each pass through alternate vertices of the other. (*Hint:* Consider the pentagons $P_{12}P_{23}P_{34}P_{45}P_{51}$ and $P_{31}P_{14}P_{42}P_{25}P_{53}$ of Ex. 3.)

2·3 The principle of duality.

The principle of duality (in two dimensions) asserts that every definition remains significant, and every theorem remains true, when we interchange

point and *line*,

join and *intersection*.

Thus the dual of $AB \cdot CD$ is $(a \cdot b)(c \cdot d)$, Axiom 2·21 is self-dual, and the dual of 2·24 is part of 2·23. Some other changes of wording are obvious consequences of these fundamental changes; e.g. the dual of 1·52 is 2·26.

To establish this principle we merely have to observe that *the axioms imply their own duals* (see §2·2, Ex. 1). Given a theorem and its proof, we can immediately assert the dual theorem; for a proof of the latter could be written down mechanically by dualizing every step in the proof of the original theorem.

Although the closely related idea of reciprocal polyhedra had already occurred in the writings of the medieval Italian Maurolycus (1494–1575), the principle of duality may properly be ascribed to Gergonne (1771–1859). Poncelet protested that it was nothing but his method of reciprocation with respect to a conic (polarity), and Gergonne replied that the conic is irrelevant— duality is intrinsic in the system. Thus Gergonne came nearer to realizing how the principle rests on the symmetrical nature of the axioms of incidence. It is sad that such a beautiful discovery was marred by bitter controversy over the question of priority.

Instead of using axioms that merely *imply* their duals, it might perhaps be more satisfactory to use an inherently self-dual set of axioms, such as the following:*

2·31   *Any two* $\begin{matrix} points \\ lines \end{matrix}$ *are incident with at least one* $\begin{matrix} line \\ point. \end{matrix}$

2·32   *Two distinct points cannot both be incident with two distinct lines.*

2·33   *There exist two points and two lines such that each of the points is incident with just one of the lines.*

2·34   *There exist two points and two lines (the points not incident with the lines) such that the join of the points is incident with the intersection of the lines.*

2·35   *If four points O, P, Q, R having six distinct joins and four lines o, p, q, r having six distinct intersections are so situated that the five joins OP, OQ, OR, PR, QR are incident with the respective intersections* $q \cdot r, r \cdot p, p \cdot q, q \cdot o, o \cdot p$, *then the sixth join PQ is incident with the sixth intersection* $o \cdot r$.

EXERCISES

1. Express 2·26 in terms of lines p, q, r, p', q', r', so as to make it formally dual to 2·25 (see Fig. 2·2B).

2. Verify that Axioms 2·21–2·25 imply 2·31–2·35, and vice versa.

---

* Axions 2·31–2·34 were kindly supplied by Karl Menger. For 2·35, see Veblen and Young 1910, p. 53, Ex. 8.

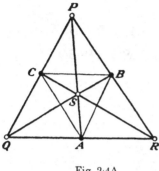

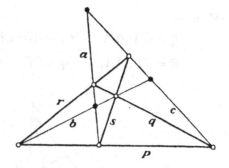

Fig. 2·4A   Fig. 2·4B

2·4   Quadrangle and quadrilateral. The following definitions are written in parallel columns to emphasize the principle of duality:

Four points P, Q, R, S, of which no three are collinear, are the vertices of a complete *quadrangle** $PQRS$, of which the six sides are the lines QR, PS, RP, QS, PQ, RS. The intersections of 'opposite' sides, namely,

$$A = QR \cdot PS,$$

$$B = RP \cdot QS,$$

$$C = PQ \cdot RS,$$

are called *diagonal points* and are the vertices of the *diagonal triangle* (see Fig. 2·4A).

Four lines p, q, r, s, of which no three are concurrent, are the sides of a complete *quadrilateral** $pqrs$, of which the six vertices are the points $q \cdot r$, $p \cdot s$, $r \cdot p$, $q \cdot s$, $p \cdot q$, $r \cdot s$. The joins of 'opposite' vertices, namely,

$$a = (q \cdot r)(p \cdot s),$$

$$b = (r \cdot p)(q \cdot s),$$

$$c = (p \cdot q)(r \cdot s),$$

are called *diagonal lines* and are the sides of the *diagonal triangle* (see Fig. 2·4B).

2·41   *If ABC is the diagonal triangle of a quadrangle PQRS, the three points*

$$A_1 = BC \cdot QR, \qquad B_1 = CA \cdot RP, \qquad C_1 = AB \cdot PQ$$

are collinear.

Proof: Apply Desargues's theorem (our Theoem 1·52 or Axiom 2·25) to the two triangles ABC and PQR (see Fig. 2·4C).

As a corollary we have

2·42   *Given the diagonal triangle and one vertex of a quadrangle, the remaining three vertices may be constructed by incidences.*

In fact, given the diagonal triangle ABC and vertex P, we construct,

---

* When there is no danger of confusion, we shall omit the word *complete*.

in turn,

$$B_1 = CA \cdot BP, \qquad C_1 = AB \cdot CP, \qquad A_1 = BC \cdot B_1 C_1,$$
$$R = BP \cdot AA_1, \qquad Q = CP \cdot AA_1, \qquad S = AP \cdot BQ.$$

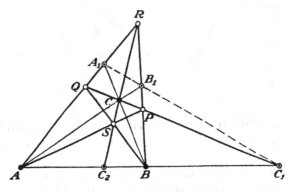

Fig. 2·4C

EXERCISES

1. Prove that the six sides of a quadrangle meet the three sides of its diagonal triangle in the six vertices of a quadrilateral which has the same diagonal triangle. (*Hint:* Define

$$A_2 = BC \cdot PS, \qquad B_2 = CA \cdot QS, \qquad C_2 = AB \cdot RS,$$

and use triangles ABC and SRQ to prove A_1, B_2, C_2 collinear.)

2. Dualize 2·41 and 2·42.

3. Show that the 10 points and 10 lines of the Desargues configuration (Fig. 2·2A or B) may be regarded (in five ways) as consisting of a quadrangle and a quadrilateral so situated that the six sides of the quadrangle pass through the six vertices of the quadrilateral (cf. 2·35).

2·5 Harmonic conjugacy. Although harmonic conjugates were used by Desargues, the following construction for them seems to have been first given by another Frenchman, La Hire (1640–1718):

Four collinear points A, B, C, D are said to form a *harmonic set* if there is a quadrangle of which two opposite sides pass through A and two other opposite sides through B, while the remaining sides pass

Four concurrent lines a, b, c, d are said to form a *harmonic set* if there is a quadrilateral of which two opposite vertices lie on a and two other opposite vertices on b, while the remaining vertices lie on c and d, through C and D, respectively. We say that C and D are *harmonic conjugates* (of each other) with respect to A and B, and we write

$$H(AB, CD)$$

as an abbreviated statement of this relation.

To construct D, given A, B, C, we draw any triangle PQR whose sides QR, RP, PQ pass through A, B, C, respectively. This determines a quadrangle $PQRS$, where

$$S = AP \cdot BQ,$$

as in Fig. 2·5A. We thus obtain

$$D = RS \cdot AB.$$

respectively. We say that c and d are *harmonic conjugates* (of each other) with respect to a and b, and we write

$$H(ab, cd)$$

as an abbreviated statement of this relation.

To construct d, given a, b, c, we draw any triangle pqr whose vertices $q \cdot r$, $r \cdot p$, $p \cdot q$ lie on a, b, c, respectively. This determines a quadrilateral $pqrs$, where

$$s = (a \cdot p)(b \cdot q),$$

as in Fig. 2·5B. We thus obtain

$$d = (r \cdot s)(a \cdot b).$$

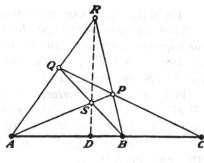

Fig. 2·5A

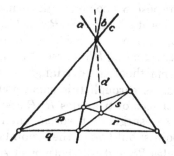

Fig. 2·5B

The construction for D or d involves the choice of a triangle. How do we know that a different triangle will lead to the same final result? This question of the *uniqueness* of the harmonic conjugate can be answered affirmatively with the help of §2·2. We need only consider the case of a harmonic set of points; that of a harmonic set of lines will follow by duality (because we know that the proof could be dualized step by step).

2·51 *The harmonic conjugate of C with respect to A and B is independent of the choice of triangle PQR.*

Proof: Suppose that another such triangle $P'Q'R'$ leads to a quadrangle $P'Q'R'S'$, as in Fig. 2·5C. We have to show that RS and $R'S'$ both determine the same point D on AB. For this

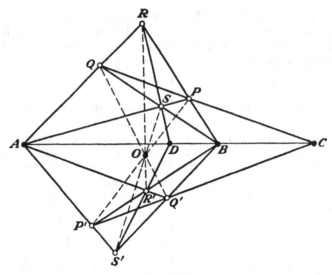

Fig. 2·5C

purpose we consider, in turn, three pairs of triangles. Corresponding sides of triangles PQR and $P'Q'R'$ meet in the three collinear points A, B, C; hence, by 2·26, the joins of corresponding vertices are concurrent, that is, RR' passes through the point $O = PP' \cdot QQ'$. Applying the same theorem to triangles PQS and $P'Q'S'$, we conclude that SS' passes through this same point O. Thus the joins of corresponding vertices of triangles RSP and $R'S'P'$ all pass through O, and hence, by Desargues's theorem, their corresponding sides meet in collinear points; but two of these points are A and B; therefore, the remaining sides RS and $R'S'$ both meet AB in the same point D.

In the notation of Fig. 2·4C, we have H$(AB, C_1 C_2)$. Thus the sides of the diagonal triangle are 'divided harmonically' by the sides of the quadrangle. Moreover,

2·52 *The sides of a quadrangle are divided harmonically by the sides of its diagonal triangle.*
   Proof: The quadrangle $PSBC$ yields H(QR, AA_1).

Incidentally, we observe that the quadrangle $CC_1 BB_1$ yields H(AA_1, QR), which is not obviously the same as H(QR, AA_1).

EXERCISES

1. Using a pencil and ruler, carry out the construction for the harmonic conjugate of C with respect to A and B, taking C somewhere between A and B. What happens if C is midway between A and B?

2. Dualize 2·51 and its proof, drawing a suitable figure.

2·6 Ranges and pencils. The points on a line are said to form a *range*, especially when we regard them as the possible positions of a *variable* point X (which runs along the line). The dual of a range is a *pencil*, consisting of the lines through one point: the possible positions of a variable line x (which rotates about the point). The common point of the lines is called the *centre* of the pencil.

We proceed to define a *correspondence* (strictly, a one-to-one correspondence) between two ranges. This is a rule for associating every point X of the first range with every point X' of the second, so that there is exactly one X' for each X and exactly one X for each X'. It is usually desirable to think of the correspondence as being directed from X to X', that is, to distinguish between this and the *inverse* correspondence from X' to X. The two ranges need not be on distinct lines. One trivial case, which must not be ignored, is when X' continually coincides with X; this correspondence is called the *identity*.

There is a similar definition for a correspondence between two pencils or between a pencil and a range. The simplest correspondence of the latter type occurs when we take the section of the pencil by a fixed line o, so that each line x of the pencil is associated with the point $X = o \cdot x$ of the corresponding range, as in Fig. 2·6A. The inverse correspondence occurs when we project a range from a fixed point O so that each point X of the range is associated with the line $x = OX$ of the corresponding pencil. The existence of these simple correspondences is one of the basic reasons for the efficacy of projective geometry.

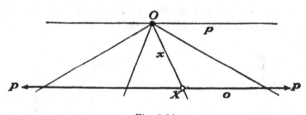

Fig. 2·6A

In affine (or Euclidean) geometry, the line p (through O) parallel to o would be exceptional, for it would have no corresponding point on o; but when we have extended the affine plane to the projective plane, the corresponding point P is just the point at infinity on o. The line x through O, rotating continuously, determines on o the point X, which runs along to the right, say, until x is parallel to o, then immediately reappears far away on the left and continues running to the right. In affine geometry the point X makes an infinite jump; but in projective geometry its motion, through the single point at infinity, is continuous.

2·7 Perspectivity. Apart from the identity, the simplest correspondence between two ranges is that which occurs when we compare

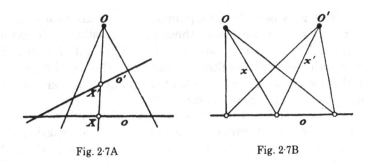

Fig. 2·7A   Fig. 2·7B

the sections of a pencil by two distinct lines o and o', as in Fig. 2·7A. The relation between X on o and X' on o' is such that the line XX' passes through a fixed point O, and we call the correspondence a *perspectivity* from O, writing

$$X \overset{o}{\barwedge} X', \quad \text{or simply} \quad X \barwedge X'.$$

Dually (Fig. 2·7B) a perspectivity from a line o occurs when the relation between two pencils is such that the point of intersection $x \cdot x'$ lies on a fixed line o; then we write

$$x \overset{o}{\barwedge} x' \quad \text{or} \quad x \barwedge x'.$$

The following important theorems illustrate the way this notation may be used:*

2·71 *It is possible, by a sequence of three perspectivities, to interchange pairs among any four collinear points.*

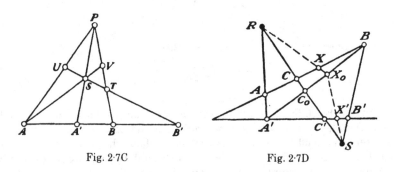

Fig. 2·7C   Fig. 2·7D

Proof: Suppose we wish to interchange A with A' and B with B', that is, to make the permutation $AA'BB' \to A'AB'B$ or $(AA')(BB')$. Draw any triangle TUP whose sides UP, PT, TU pass through A, B, B'. This

---

* von Staudt 1847, p. 59, § 119. This use of the symbol $\barwedge$ is due to Veblen and Young 1910, p. 57; von Staudt used it differently.

determines two further points

$$S = A'P \cdot TU, \qquad V = AS \cdot BP$$

as in Fig. 2·7C, and we have

$$AA'BB' \overset{P}{\overline{\wedge}} USTB' \overset{A}{\overline{\wedge}} PVTB \overset{S}{\overline{\wedge}} A'AB'B.$$

2·72 *It is possible, by a sequence of not more than three perspectivities, to relate any three distinct collinear points to any other three distinct collinear points.*

Proof: If the two triads, ABC and $A'B'C'$, are on distinct lines, as in Fig. 2·7D, let R, S, C_0 denote the points where the respective lines AA', BB', BA' meet CC'. Then ABC and $A'B'C'$ are related by the sequence of two perspectivities

$$ABC \overset{R}{\overline{\wedge}} A'BC_0 \overset{S}{\overline{\wedge}} A'B'C'.$$

(If A and A' coincide, we merely use the perspectivity from S.)

If the two triads are on one line, we use a quite arbitrary perspectivity $ABC \overline{\wedge} A_1B_1C_1$ to obtain a triad on another line and then relate $A_1B_1C_1$ to $A'B'C'$ by the above construction.

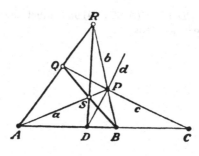

Fig. 2·8A

2·8 The invariance and symmetry of the harmonic relation.
We proceed to show that harmonic sets remain harmonic after any number of perspectives. As a first step we shall prove

2·81 *Any section of a harmonic set of lines is a harmonic set of points, and a harmonic set of points is projected from any point by a harmonic set of lines.*

Remark: This theorem is in two dual parts, and thus it will suffice to prove the latter part: If A, B, C, D are joined to a point P (outside their line) by lines a, b, c, d and if H(AB, CD), then H(ab, cd).

Proof: Let P be used as a vertex of the triangle PQR in constructing D from A, B, C, as in Fig. 2·5A or 2·8A. Then the quadrilateral

$ASBRQD$ has two opposite vertices on $AS = a$, two others on $BR = b$, one vertex Q on c, and one vertex D on d. Hence H(ab, cd).

Combining the two parts of 2·81, we conclude that

2·82 *Perspectivities preserve the harmonic relation:*
If $ABCD \barwedge A'B'C'D'$ and H(AB, CD), then H($A'B'$, $C'D'$).

Our definition for harmonic conjugacy involves A and B symmetrically and likewise C and D. Hence

2·83 *The four relations*

$$H(AB, CD), \quad H(BA, CD), \quad H(AB, DC), \quad H(BA, DC)$$

are all equivalent.

EXERCISES

1. Show that Fig. 2·8A can be made formally self-dual by interchanging the names of c and d (so that $c = PD$ and $d = PC$). Which lines should then be named p, q, r, s?

2. If we have H(AB, CD) and H($A'B'$, $C'D$) on distinct lines, show that the three lines AA', BB', CC' are concurrent.

CHAPTER 3

Order and Continuity

The order of arrangement of lines in a pencil, like that of points on a circle, is cyclic; we cannot say of three that one is between the other two, but we can say of four that two *separate* the other two. The correspondence between a pencil and its section enables us to carry over this cyclic order from pencils to ranges. If A and B separate C and D, we write $AB//CD$. (The idea of a point C lying *between* A and B belongs to affine geometry and may be interpreted as meaning that $AB//CD$, where D is the point at infinity on AB.)

The basic properties of separation may be stated in the form of six axioms, as in § 3·1. These are not quite sufficient for a complete characterization of the real projective line. The final axiom, concerning continuity, will be introduced in § 3·5. This particular form has been chosen because it is ready for immediate application in proving the fundamental theorem of projective geometry and other theorems; moreover, it is intuitively acceptable. Dedekind's axiom has not been used, for it is more difficult to grasp and to apply. The deduction of our axiom from Dedekind's was carried out by Enriques.[*] In Chapter 10 we shall consider a third possible approach to the theory of continuity.

3·1 The axioms of order.

3·11 *There exists a line containing four distinct points.*

3·12 *If $AB//CD$, then A, B, C, D are four distinct collinear points.*

3·13 *If $AB//CD$, then $AB//DC$.*

3·14 *If A, B, C, D are four distinct collinear points, at least one of the three relations $BC//AD$, $CA//BD$, $AB//CD$ must hold.*

---

[*] Enriques 1930, pp. 71–5, Cf. § 10·6.

3·15   *If AB//CD and AC//BE, then AB//DE.*

3·16   *If AB//CD and ABCD $\overline{\wedge}$ A'B'C'D', then A'B'//C'D'.*

The first five of these six axioms have been adapted from those given by Vailati. They express obvious properties of points arranged around a circle. Accordingly, we often find it convenient to use a circular diagram when dealing with points on a single line; e.g. Fig. 3·1A illustrates 3·15. (This does not mean that we imagine the line to be somehow 'bent', but it emphasizes the important fact that the line is 'closed'.)

In contrast to these five one-dimensional axioms, 3·16 is essentially two-dimensional, relating order on one line to order on another. It enables us to derive the dual statements concerning the relation *ab//cd* for concurrent lines.

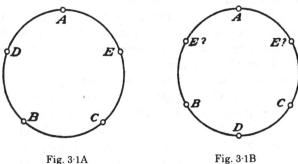

Fig. 3·1A   Fig. 3·1B

Using 2·71 to interchange the pairs (*AC*)(*BD*), we deduce from 3·16 that

$$AB//CD \quad \text{implies} \quad CD//AB.$$

Taking this with 3·13, we conclude that the eight relations

$$AB//CD, \quad BA//CD, \quad AB//DC, \quad BA//DC,$$

$$CD//AB, \quad CD//BA, \quad DC//AB, \quad DC//BA,$$

are all equivalent (cf. 2·83).

By 3·12, the *D* and *E* of 3·15 cannot coincide. Thus the relation *AB//CD* excludes *AC//BD*; but these are equivalent to two of the relations in 3·14, and we can argue similarly for any other two. Hence

3·17   *The three relations BC//AD, CA//BD, AB//CD are mutually exclusive—no two can hold simultaneously.*

The following theorem has a somewhat similar appearance:

3·18   *The three relations BC//DE, CA//DE, AB//DE cannot all hold simultaneously.*

Proof*: Assuming all three, we shall find that each of the relations in 3·14 leads to a contradiction. Suppose, for instance, that $BC//AD$ (or $AD//BC$), as in Fig. 3·1B. Then by 3·15 the relations $AD//BC$ and $AB//DE$ imply $AD//CE$, which excludes $CA//DE$ (by 3·17). Since A, B, C enter the discussion symmetrically, the proof is complete.

This tricky proof of an obvious statement indicates the economy of our axioms; they are only just strong enough to provide the familiar properties of cyclic order.

From 2·72, 3·11, 3·14, and 3·16 we deduce

3·19 *If A, B, C are three distinct collinear points, their line contains a point D such that $AB//CD$.*

EXERCISES

1. Prove that $AB//CD$ and $AC//BE$ imply $BE//CD$. (*Hint: $BE//AC$ and $BA//ED$.*)

2. Show that 3·15 can be replaced by the following, more readily memorable, axiom (Wyler 1952, p. 657):
If $AB//CD$ and $BC//DE$, then $CD//EA$.

3·2 Sense. If A, B, C are three distinct collinear points, we define the *segment AB/C†* as consisting of all points X for which $AB//CX$. (Thus the segment AB/C does *not* contain C. The familiar segment AB of affine geometry may be described as AB/C with C at infinity.) The segment plus its end points A and B is called an *interval* and written $\overline{AB}/C$. If X and Y belong to $\overline{AB}/C$, the interval $\overline{XY}/C$ is said to be *interior* to $\overline{AB}/C$ (even if X or Y coincides with A or B), and a point D lies *between* X and Y in $\overline{AB}/C$ if it belongs to XY/C, that is, if $XY//CD$. Thus the notion of intermediacy (or three-point order) is valid for an interval, although not for the whole line. In a circular diagram such as Fig. 3·1A an interval naturally appears as an arc.

In ordinary (affine) geometry a point decomposes a line through it into two rays; but in projective geometry a point does not divide a line at all. (We can reach the left side of the barrier point from the right by proceeding to the right and passing through the point at infinity.) The circular diagram suggests that two points will decompose their joining line into just two segments (represented by two semicircles, say); but this is quite difficult to prove, as we shall see.

3·21 *If $AB//CD$, the two points A and B decompose their line into just two segments: AB/C and AB/D.*

---

* Due to N.D. Lane and F.A. Sherk.

† Read as 'AB without C'.

Proof: No point X can lie in both AB/C and AB/D; for then we should have $XC//AB$, $CD//AB$, $DX//AB$, which are incompatible (by 3·18). It remains to be shown that any point X, other than A or B, must lie in *one* of the segments.* We shall assume that X does not lie in AB/C and deduce that it then lies in AB/D. This is clearly true when X coincides with C. By 3·14 (with X for D), the only other possibility is either $BC//AX$ or $CA//BX$. In the former case we have $BA//CD$ and $BC//AX$, implying $BA//DX$; in the latter, $AB//CD$ and $AC//BX$, implying $AB//DX$. Thus in either case X belongs to AB/D, as required.

Two such segments, and likewise the corresponding intervals $\overline{AB}/C$ and $\overline{AB}/D$, are said to be *supplementary*.

Axiom 3·14 and Theorem 3·17 imply that three collinear points A, B, C decompose their line into three segments BC/A, CA/B, AB/C; it can be proved by induction that n collinear points decompose their line into n segments.† More precisely, n distinct collinear points can be named P_0, P_1, ..., P_{n-1} (with subscripts reduced modulo n, so that $P_n = P_0$, $P_{n+1} = P_1$, and so on) in such an order that the n segments into which they decompose their line are $P_{r-1}P_r/P_{r+1}$, where r takes each of the n possible values in turn. There are $2n$ ways in which these symbols can be assigned to the n points: any one point may be called P_0, and its neighbours are then P_1 and P_{n-1} or *vice versa*. Thus the line contains infinitely many points.

In this manner the notion of the cyclic order of any number of points may be rigorously justified, but it would be tedious to give all the logical details here. Enough has been said to show the reader that he can henceforth safely rely on intuition, in the certainty that he could, if he wished, supply the proofs.

One consequence of this notion of cyclic order is the distinction of *sense*.‡ In Fig. 3·1A the sense ABC is 'positive' or 'counterclockwise' while ABD (or BDA or DAB) is 'negative' or 'clockwise'. In Fig. 2·5A the sense ABC is 'left to right' while ABD is 'right to left'. To analyse these notions, let ABC and DEF be two sets of three distinct points on the same line. We write

$$S(ABC) = S(DEF) \quad \text{or} \quad S(ABC) \neq S(DEF)$$

according as the sense DEF agrees or disagrees with the sense ABC. Since any of D, E, F may coincide with any of A, B, C, we are now dealing with n distinct collinear points where n may be 3, 4, 5, or 6. Let these points be renamed P_0, P_1, ..., P_{n-1} in such an order that A, B, C are P_0, P_b, P_c with $0 < b < c < n$. Suppose the symbols thus determined for D, E, F are P_d, P_e, P_f. The distinction of sense may be *defined* by

---

* This part of the proof is due to Robinson 1940, p. 120.

† Veblen and Young 1918, p. 46.

‡ Coxeter 1967, p. 32.

writing

$$S(ABC) = S(DEF) \quad \text{if } d < e < f \text{ or } e < f < d \text{ or } f < d < e,$$

$$S(ABC) \neq S(DEF) \quad \text{if } f < e < d \text{ or } d < f < e \text{ or } e < d < f.$$

We easily verify (by taking $n = 3$) that

$$S(ABC) = S(BCA) = S(CAB) \neq S(CBA).$$

Similarly (with $n = 4$ and $A = P_0,\ D = P_1,\ B = P_2,\ C = P_3$),

3·22 *The relation $AB//CD$ is equivalent to $S(ABC) \neq S(ABD)$.*

Instead of regarding separation as a primitive concept, Veblen and Young (1918, p. 32) considered 'undefined elements called senses' and derived separation by using 3·22 as a definition. It seems reasonable to argue that separation is a simpler concept than sense, because the relation $AB//CD$ involves only four points whereas $S(ABC) = S(DEF)$ involves six.

EXERCISE

Show that the relation $S(ABD) = S(BCD)$ implies $S(ABD) = S(ACD)$.

3·3 The Sylvester-Gallai theorem. The following result, which plays a useful role in the theory of 'harmonic separation', is particularly interesting because, after its enunciation by Sylvester in 1893, it remained unproved for about forty years. Then T. Gallai proved it by an ingenious argument using parallel lines. The projective proof given here is due to R. Steinberg.*

3·31 *If n given points are not all on one line, there exists a line containing exactly two of them.*
Proof: Suppose, if possible, that every join of two of the n points contains a third. Since the points are not all collinear, they must include three forming a triangle PQR. Let p be a line through P that contains no other point of the set. All the joins of pairs of the n points meet p in a certain set of at least two points: P itself, one on QR, and possibly others. These points occur in a certain cyclic order. Let A be consecutive to P in this order, so that one of the segments AP is not met by any of the joins. This point A is not one of the n but lies on a line containing at least three of them, say B, C, D, so named that $AB//CD$. Since P and B are two of the n points, their join must contain a third, say O. Suppose $ABCD \barwedge APC'D'$. Then $AP//C'D'$; that is, the joins OC and OD each meet one of the two segments AP, contrary to our defini-

---

* For the whole story, see Borwein and Moser 1990.

tion of A. Hence, in fact, one of the joins must contain only two points of the set.

The next theorem, which we tacitly assumed in § 2·5, is sometimes taken as an axiom ('Fano's axiom').*

3·32   *The three diagonal points of a quadrangle are never collinear.*
Proof: A quadrangle with collinear diagonal points would constitute a configuration of seven points and seven lines such that every two of the points are joined by a line containing a third. By 3·31 with $n = 7$, this is impossible.

Considering the quadrangle $ABPQ$ of Fig. 2·5A, we deduce

3·33   *If A, B, C are all distinct*, $\mathrm{H}(AB, CD)$ *implies* $D \neq C$.

In other words, if A, B, C are distinct and $\mathrm{H}(AB, CD)$ then A, B, C, D are all distinct so that, by 3·14 and 3·17, exactly one of the three relations $BC//AD$, $AC//BD$, $AB//CD$ must hold. Since the relation $\mathrm{H}(AB, CD)$ is equivalent to $\mathrm{H}(BA, CD)$, neither of the relations $BC//AD$, $AC//BD$ could hold without the other. Hence, among the three relations, the one that holds can only be the third:†

3·34   *If A, B, C are all distinct*, $\mathrm{H}(AB, CD)$ *implies* $AB//CD$.

Combining 3·33 with 2·72 and 2·82, we deduce that

3·35 $\mathrm{H}(AB, CD)$ *implies* $\mathrm{H}(CD, AB)$, *and the four equivalent rela-*tion 2·83 can be extended to eight.

EXERCISE

Given n points, not all collinear, prove that by joining every two of them we obtain at least n distinct lines (P. Erdös). (*Hint:* Assume the result for $n - 1$ points, viz. those derived from the n by omitting one of the two whose join contains no more.)

3·4   Ordered correspondence. We have already described the concept of a correspondence between two ranges, illustrating it by the particular correspondence called perspectivity, which (by Axiom 3·16) preserves the relation of separation and consequently cyclic order and the distinction of sense. But this property of a perspectivity is shared by many other kinds of correspondence. Let us use the name *ordered*

---

* Veblen and Young 1910, p. 45; Coxeter 1969, p. 231.
† This neat proof was kindly contributed by Wilbur Jonsson.

correspondence whenever the relation of separation is preserved. That is to say, the characteistic property of an ordered correspondence $X \rightarrow X'$ is that, if the relation $AB//CD$ holds for four positions A, B, C, D of X, then the relation $A'B'//C'D'$ holds for the corresponding positions of X'. It follows that segments correspond to segments and intervals to intervals.

Any point M that coincides with its corresponding point M' is called an *invariant* point. (Some authors prefer to call it a *double* point.) For instance, a perspectivity between two ranges has just one invariant point, where the two lines intersect. Of course, there are some ordered correspondences that have no invariant points. On the other hand, there may be more than one invariant point, but in such a case it is obvious that both ranges must be on the same line.

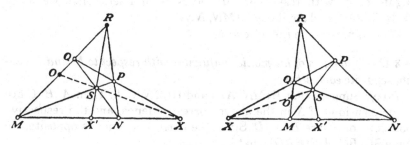

Fig. 3·4A

Any ordered correspondence preserves the distinction of sense, i.e. the relation $S(ABC) = S(DEF)$ implies $S(A'B'C') = S(D'E'F')$; but in the case of a correspondence between superposed ranges (on one line) the question arises as to whether the sense ABC agrees or disagrees with the sense $A'B'C'$. We call such a correspondence *direct* or *opposite* according as

$$S(ABC) = S(A'B'C') \quad \text{or} \quad S(ABC) \neq S(A'B'C').$$

Whichever relation holds for one triad of points must still hold for any other, in view of the above remark about $S(ABC)$ and $S(DEF)$. In particular, the identity (namely, $X \rightarrow X$) is direct.

A particularly important kind of opposite correspondence is described in the following theorem:

3·41 *The correspondence between the points of a range and their harmonic conjugates with respect to two fixed points M and N is an opposite correspondence with invariant points M and N.*

Proof: Since a perspectivity preserves order, so does the resultant or *product* of any sequence of perspectivities. To prove that the correspondence between harmonic conjugates with respect to M and N is ordered, we exhibit it as the product of three perspectivities with centres Q, M, R, in the notation of Fig. 3·4A. Here Q and R are any fixed points collinear with M. A variable point X on the line MN determines

$$P = NR \cdot QX, \quad S = MP \cdot NQ, \quad X' = MN \cdot RS,$$

and we have

$$MNX \overset{Q}{\barwedge} RNP \overset{M}{\barwedge} QNS \overset{R}{\barwedge} MNX'.$$

Thus the correspondence $X \to X'$ is ordered and has M and N as invariant points. Finally, it is opposite, since

$$\mathrm{S}(MNX) \neq \mathrm{S}(MNX')$$

by 3·22 and 3·34.

The invariance of M makes it natural to say that the harmonic conjugate of M with respect to M and N is M itself. Thus we write H(MN, MM), and similarly H(MN, NN).

We can now prove the following:

3·42 *Two pairs of harmonic conjugates with respect to M and N cannot separate each other.*

Proof: Suppose that H(MN, AB) and H(MN, CD). Then A, B, C are three positions of X in the above correspondence, and the respective positions of X' are B, A, D. Since the correspondence is opposite, we have $\mathrm{S}(ABC) \neq \mathrm{S}(BAD)$, that is,

$$\mathrm{S}(ABC) = \mathrm{S}(ABD).$$

By 3·22 this means that A and B do not separate C and D.

In proving 3·41 we exhibited the correspondence $X \to X'$, where H(MN, XX'), as the product of three perspectivities (see Fig. 3·4A). By naming one further point $O = MQ \cdot SX$ we can reduce the number of perspectivities to two. For since H(MQ, OR), O is a fourth fixed point on the line MQR, and we have

$$MNX \overset{O}{\barwedge} QNS \overset{R}{\barwedge} MNX'.$$

Interchanging N and X, as in Fig. 3·4B, we obtain a new correspondence $X \to X'$, where now H(MX, NX'):

3·43 *When* H(MX, NX'), *where M and N are fixed, $X \to X'$ is a direct correspondence with invariant points M and N.*

Proof: Using three fixed points O, Q, R (outside the line MN) such that H(MQ, OR), we observe that QX meets RX' in a point S on ON, and

$$MNX \overset{Q}{\barwedge} ONS \overset{R}{\barwedge} MNX'.$$

In this case $\mathrm{S}(MNX) = \mathrm{S}(MNX')$; for otherwise we should have

$MN//XX'$, whereas the relation H(MX, NX') implies $MX//NX'$. Thus
the correspondence is direct.

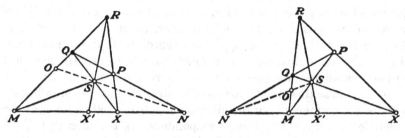

Fig. 3·4B

In ordinary analytic geometry a point on the x-axis is located by means of an
abscissa x, which measures its distance to the right of the origin (so that points
to the left of the origin have negative abscissae). A correspondence $X \rightarrow X'$ on
this axis is represented by a correspondence of abscissae:

$$x \rightarrow x' = f(x),$$

where $f(x)$ is a single-valued function such that, for any given x', the equation
$f(x) = x'$ has a unique solution. Since we are considering the *projective* line, we
must include ∞ ($= -\infty$) as a possible value for x or x'. If x and x' become
infinite together, the point at infinity is invariant; if not, there will be a finite
$x' = f(\infty)$ and a finite x for which $f(x) = \infty$. In the special case when $f(x)$ is x
itself, the correspondence is the identity. If M and N have abscissae 0 and ∞,
the correspondences considered in 3·41 and 3·43 are, respectively, $x' = -x$ and
$x' = \frac{1}{2}x$. Our axioms have been chosen so as to enable us to develop the same
theory without having recourse to analysis.

EXERCISES

1. Show that the correspondence $x \rightarrow x'$ is ordered if x' is a differentiable function
whose derivative dx'/dx never changes sign. It is direct if $dx'/dx > 0$ almost
everywhere (i.e. except where x' is infinite and possibly at some isolated places
where the derivative may vanish) and opposite if $dx'/dx < 0$ almost everywhere.
(*Hint*: Compare the signs of $(a' - b')/(a - b)$ and $(b' - c')/(b - c)$, using the
mean-value theorem.)

2. Assuming the function $x' = f(x)$ to be continuous (except where it becomes
infinite), show that the graph $y = f(x)$ is either all in one piece with no asymptote
(or with one or two oblique asymptotes) or in two pieces with two asymptotes, one
horizontal and one vertical. Which points on the graph represent invariant points
of the correspondence?

3. Show that $x \rightarrow x^3$ is a direct correspondence with four invariant points (where
$x = -1, 0, 1, \infty$).

3·5 Continuity. To get a picture of what is happening in an ordered correspondence $X \to X'$ on one line, think of a circular race track that two runners agree to run all around, starting at the same time and finishing at the same time, never stopping or turning back but otherwise free to go as fast or slow as they please. Then X and X' are the respective positions of the two runners at any instant. The correspondence is direct or opposite according as the runners are going in the same direction or in opposite directions. An invariant point occurs where the runners meet or where one overtakes the other. In the direct case this may happen any number of times, even infinitely often, for the runners might remain side by side for awhile (or even for the whole journey, when the correspondence is the identity). Thus there may be any number of invariant points, from none at all to infinitely many. But in the opposite case a little thought reveals that the runners will meet exactly twice before each returns to his own starting point (or if they started from the same point, they will meet once more elsewhere). This means that every opposite correspondence should have exactly two invariant points; but we cannot prove this rigorously without introducing one further assumption, such as the following:

3·51 Axiom of continuity. *If an ordered correspondence relates an interval* $\overline{AB}/C$ to an interior interval $\overline{A'B'}/C$, then the latter contains an invariant point M such that there is no invariant point between A and M (in $\overline{AB}/C$).*

This is obvious when we think of the race-track: X runs from A to B while X' runs *over part of the same ground* from A' to B'; M is the first point where they meet.

If the correspondence is opposite, the last clause of the axiom (after 'point M') is superfluous; M is the *only* invariant point in $\overline{AB}/C$. For, two invariant points (such as the M and N of Fig. 3·5A) would determine a segment whose sense is preserved.

3·6 Invariant points. We are now ready to prove the following:

3·61 *Every opposite correspondence has exactly two invariant points.*
Proof: Since the identity is direct, any opposite correspondence admits a point A that is not invariant. Suppose the correspondence relates A to A' and A' to A''. Choose a point C such that $AA'//A''C$ (or, if A'' coincides with A, take *any* new point C). Then the given opposite correspondence relates $\overline{AA'}/C$ to the interior interval $\overline{A'A''}/C$, as in

---

* We use the interval, rather than the segment AB/C, to cover the possibility of B' coinciding with B, in which case M might also coincide with B.

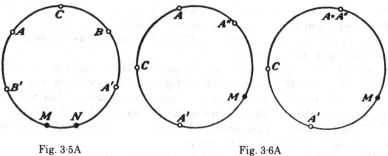

Fig. 3·5A   Fig. 3·6A

Fig. 3·6A. Hence there is just one invariant point M in AA'/C. Similarly, there is a second invariant point N in the supplementary segment AA'/M; for the inverse correspondence (namely, $X' \to X$) relates $\overline{A'A''}/M$ to the interior interval $\overline{A'A'}/M$.

We saw, in 3·42, that the relations

$$H(MN, AB) \quad \text{and} \quad H(MN, CD)$$

preclude $AB//CD$. This theorem has an important converse:

3·62 *If AB and CD are two pairs of points that are collinear but do not separate each other, then there exist points M and N such that*

$$H(AB, MN) \quad and \quad H(CD, MN).$$

Proof: Any point X has a harmonic conjugate X^I with respect to A and B and a harmonic conjugate X^J with respect to C and D—in symbols,

$$H(AB, XX^I) \quad \text{and} \quad H(CD, XX^J).$$

While X runs from A to B over the interval $\overline{AB}/C$, X^I runs from A to B over the supplementary interval $\overline{AB}/C^I$, which includes D as well as C, since the pairs AB and CD do not separate each other (see Fig. 3·6B). Meanwhile, X^J runs from A^J to B^J over part of the same interval. Now consider the combined correspondence $X^I \to X^J$. This relates the interval $\overline{AB}/C^I$ to the interior interval $\overline{A^J B^J}/C$. By Axiom 3·51, the latter interval contains an invariant point M, which can equally well be called N^I or N^J, since it is the harmonic conjugate of some point N with respect to either of the pairs AB, CD (see Fig. 3·6C).

The theorem we have just proved is especially significant, since it would enable us to define separation in terms of incidence, instead of taking separation to be a second undefined relation. In fact, we could define $AB//CD$ to mean that there is no common pair of harmonic conjugates with respect to the two pairs AB and CD. Then we could prove 3·12, 3·13, and 3·16 as theorems, leaving only three axioms of order. This idea is due to Pieri, whose exposition was praised by Russell* in the following words: 'This is, in my opinion, the best work on the present subject.' Actually, instead of AB/C, Pieri defined the supplementary segment (ACB), containing C; and he took 3·32 as an axiom. His definition† may be expressed thus:

(ACB) *is the locus of the harmonic conjugate of* C *with respect to a variable pair of distinct points that are harmonic conjugates with respect to* A *to* B.

Adding to the segment its end points A and B, we obtain an interval, say $(\overline{A}C\overline{B})$; and the relation $AB//CD$ means that D, on AB, does not belong to $(\overline{A}C\overline{B})$. Thus Pieri reduced the undefined relations to incidence alone and reduced the axioms of order to the following three:

(1) *If* D, *on* AB, *does not belong to* $(\overline{A}C\overline{B})$, *it belongs to* (ABC).

(2) *If* D *belongs to both* (ABC) *and* (BAC), *it cannot belong to* (ACB).

(3) *If* D *belongs to* (ACB) *and* E *to* (ADB), *then* E *belongs to* (ACB).

On the other hand, this simplification is to some extent illusory, as these axioms would be quite complicated if we expressed them directly in terms of incidence. Now, which is preferable: a number of simple axioms involving two undefined relations, or fewer but far more complicated axioms involving only one such relation? The answer is a matter of taste.

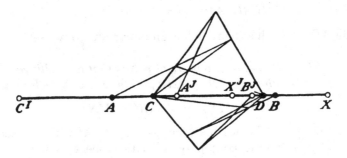

Fig. 3·6B

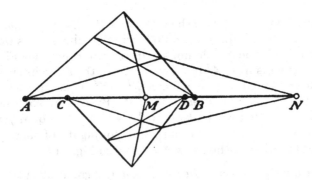

Fig. 3·6C

---

* 1930, p. 382.
† 1899, p. 24.

EXERCISES

1. Prove that if D belongs to (ACB), $(ACB) = (ADB)$. (*Hint:* By Pieri's definition of a segment, C belongs to (ADB). By (3), every point of (ADB) belongs to (ACB). By the same axiom with C and D interchanged, every point of (ACB) belongs to (ADB).)

2. Deduce 3·11, 3·14, and 3·15 from Pieri's axioms. (*Hint:* To deduce 3·11, remember that Pieri's axioms include 3·32. As for 3·14, this is a simple restatement of (2). To prove 3·15 we may argue as follows.* Since $AB//CD$ and $AC//BE$, D does not belong to $(\overline{A}C\overline{B})$, nor E to $(\overline{A}B\overline{C})$. Hence by (1), E belongs to (ACB) while D does not. Interchanging D and E in (3), we conclude that D cannot belong to (AEB), nor even to $(\overline{A}E\overline{B})$. Hence $AB//DE$.)

3·7   Order in a pencil. If a, b, c, d are four concurrent lines meeting another line in points A, B, C, D such that $AB//CD$, then we say $ab//cd$. By Axiom 3·16 this definition for separation of line pairs is independent of the chosen section $ABCD$. We can easily dualize all the results of the present chapter; e.g. the dual of a segment is an angle. If a, b, c are three lines through a point O, we can distinguish the two senses of rotation about O as $S(abc)$ and $S(cba)$. We can define an *ordered correspondence* between two pencils. If the pencils have the same centre, the correspondence may be *direct* or *opposite*. By 3·61 every opposite correspondence between pencils has exactly two invariant lines.

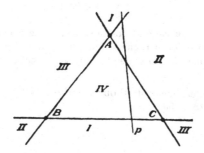

Fig. 3·8A

3·8   The four regions determined by a triangle. We saw, in 3·21, that two points decompose their line into just two segments. Dually, two lines decompose the pencil to which they belong into two angles, or we may say that they decompose the whole plane into two angular *regions*. ('Vertically opposite' angles belong to the same region, since

---

* This solution was given by R.G.E. Epple when he was a graduate student at the University of Southern California.

the line at infinity forms no barrier between them.) A third line, not concurrent with the first two, penetrates both regions. Hence:

3·81 *Three lines that form a triangle decompose the whole plane into four regions.* *

A fourth line, not through a vertex of the triangle, is decomposed by the three lines into three segments, one in each of some three of the four regions. Hence there is just one of the four regions that the new line fails to penetrate. This suggests a notation for distinguishing the four regions formed by lines *BC, CA, AB*; the region *not* penetrated by a line *p* is denoted by *ABC/p*. (In Fig. 3·8A, this is region *III*.)

EXERCISES

1. Observe that the interior of the ordinary triangle *ABC* of affine geometry may be described as *ABC/o*, where *o* is the line at infinity.

2. In Euclidean geometry, three lines forming a triangle are tangents to four circles (inscribed and escribed), one in each of the four regions.

3. In ordinary analytic geometry the coordinate axes *OX, OY* and the line at infinity may be regarded as forming a triangle that decomposes the plane into the four *quadrants*. Show that each of these may be denoted by *OXY/p*, where *p* is one of the four lines

$$\pm x \pm y + 1 = 0.$$

4. Show that four lines of general position (forming a complete quadrilateral) decompose the plane into seven regions: four triangular (bounded by three segments) and three quadrangular (bounded by four segments).

---

* For a more extended account (though without the notation *ABC/p*) see Veblen and Young 1918, p. 53.

CHAPTER 4

One-Dimensional Projectivities

The present chapter is concerned with the most important kind of ordered correspondence: the projectivity, which may be defined either as the product of several perspectivities or as a correspondence that preserves harmonic sets. The first definition, due to Poncelet, has been adopted by Veblen, Baker, and other authors; it has the advantage of remaining valid in complex geometry. This book, however, follows Enriques in using the second definition, due to von Staudt, which generalizes more readily to two (or more) dimensions. It is an immediate consequence of 2·82 that every Poncelet projectivity is a von Staudt projectivity, and we shall prove in §4·2 that every von Staudt projectivity (in real geometry) is a Poncelet projectivity. Thus from that point on the two treatments coincide.

4·1   Projectivity. The notion of correspondence extends easily from the line to the plane. By a two-dimensional correspondence $X \to X'$ we mean a rule for associating every point X with every point X' so that there is exactly one X' for each X and exactly one X for each X'. A correspondence between lines, $x \to x'$, is defined similarly.*

A *collineation* is the special case where collinear points correspond to collinear points, and consequently concurrent lines to concurrent lines; i.e. ranges correspond to ranges, and pencils to pencils. Thus a collineation preserves incidences: point X' lies on line x' if and only if point X lies on line x. The range of points X on a given line x corresponds to a range of points X' on the corresponding line x'. Four positions of X forming a harmonic set correspond to four positions of X' forming a harmonic set; for any quadrangle used in constructing the first set corresponds to a quadrangle having the same relation to the second set. This suggests the following one-dimensional analogue:

---

* If we had not restricted our geometry to two dimensions (by means of Axiom 2·24), we could just as easily have defined a correspondence between the points (or lines) of two distinct planes.

A *projectivity* between two ranges is a correspondence that preserves the harmonic relation. In other words, if the relation H(AB, CD) holds for four positions A, B, C, D of X, then the relation H($A'B'$, $C'D'$) holds for the corresponding positions of X'. The established notation, invented by von Staudt,* is

$$X \barwedge X'.$$

Thus the relations $ABCD \barwedge A'B'C'D'$ and H(AB, CD) imply H($A'B'$, $C'D'$).

By 2·82 the perspectivity $X \overline{\overline{\wedge}} X'$ is a special case of the projectivity $X \barwedge X'$. We may now write 2·71 in the concise form

$$ABCD \barwedge BADC \barwedge CDAB \barwedge DCBA$$

(for any four collinear points).

We also define a projectivity $x \barwedge x'$ between the lines of two pencils: if H(ab, cd) holds for four positions a, b, c, d of x in the first pencil, then H($a'b'$, $c'd'$) holds for the corresponding four positions of x' in the second.

The following theorem will enable us to apply to projectivities some of the results already obtained for ordered correspondences (e.g. 3·61):

4·11 *Every projectivity is an ordered correspondence.* In other words, if $ABCD \barwedge A'B'C'D'$ and $AB//CD$, then $A'B'//C'D'$.

Proof: Suppose, if possible, that $ABCD \barwedge A'B'C'D'$ and $AB//CD$ but not $A'B'//C'D'$. Then by 3·62 there exist points M' and N' such that H($A'B'$, $M'N'$) and H($C'D'$, $M'N'$). These two points of the second range correspond to points M and N of the first, such that H(AB, MN) and H(CD, MN). By 2·83 this means that H(MN, AB) and H(MN, CD). But we have assumed $AB//CD$; thus 3·42 is contradicted.

The next theorem shows a radical departure from the general ordered correspondence (which, if direct, may have any number of invariant points):

4·12 *A projectivity having more than two invariant points can only be the identity.*†

Proof: We shall obtain a contradiction by supposing that a given projectivity has three invariant points A, B, C and a noninvariant point P, so that $ABCP \barwedge ABCP'$ with $P \neq P'$. Let the points A, B, C be named in such an order that P lies in the segment AB/C and P' in PB/C (see Fig. 4·1A). The projectivity relates the interval $\overline{PB}/C$ to the interior interval $\overline{P'B}/C$. Hence by Axiom 3·51 the latter interval contains a 'first' invariant point M (admitting no invariant points between P and

---

* 1847, p. 49.
† von Staudt 1847, p. 50, § 106.

M). Similarly the inverse projectivity $(X' \barwedge X)$ relates $\overline{AP'}/C$ to $\overline{AP}/C$, which consequently contains a 'last' invariant point N (admitting no invariant points between N and P'). Since the segments NP'/C and $^DM/C$ overlap, we can assert that the segment NM/C is entirely free t·om invariant points.

Let D be the harmonic conjugate of C with respect to M and N, and suppose $D \barwedge D'$. Since $MNCD \barwedge MNCD'$, the relation H(MN, CD) implies H(MN, CD'). Hence, by 2·51, $D = D'$, and D is an invariant point in the forbidden segment MN/C, which is absurd. Thus there cannot really be three invariant points (unless *every* point is invariant).

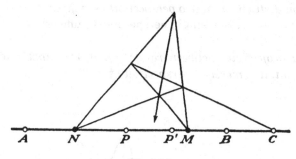

Fig. 4·1A

4·2 The fundamental theorem of projective geometry. The following theorem derives its name from the fact that it opens the way to the most characteristic developments of our subject. Its strength will be seen in the ease with which the remaining theorems of this chapter can be proved. Moreover, it enables us to construct any given projectivity as a product of perspectivities, thus reconciling the treatments of Poncelet and von Staudt.

4·21 The fundamental theorem. *A projectivity is determined when three points of one range and the corresponding three points of the other are given.*[*]

Proof: Suppose we are given three points A, B, C of one range and corresponding points A', B', C' of the other. We proceed, as in the proof of 2·72 (Fig. 2·7D), to construct a projectivity $X \barwedge X'$ such that $ABC \barwedge A'B'C'$.

To establish the uniqueness of this projectivity, we have to prove that a different construction (e.g. by joining AB' instead of BA') would yield the same X' for a given X. Suppose one construction gives

---

* von Staudt 1847, p. 52, § 110.

$$ABCX \barwedge A'B'C'X',$$

while another gives

$$ABCX \barwedge A'B'C'X_1'.$$

Then by combining the two constructions we obtain

$$A'B'C'X' \barwedge A'B'C'X_1'.$$

This combined projectivity has three invariant points A', B', C'; hence, by 4·12, X_1' must coincide with X'.

4·22   Corollary. *Any projectivity can be constructed as a product of perspectivities, the number of which can be reduced to three. If the two ranges are on distinct lines, two perspectivities suffice.**

In one important case a single perspectivity suffices:

4·23   *If a projectivity between ranges on two distinct lines has an invariant point, it is merely a perspectivity.*†

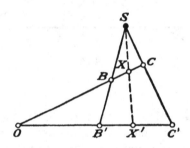

Fig. 4·2A.

Proof: Of course, the invariant point O belongs to both ranges; thus it must be the point of intersection of the two lines, as in Fig. 4·2A. Let B and C be any other points of the first range, B' and C' the corresponding points of the second. Then we have $OBC \barwedge OB'C'$. But

$$OBC \overset{S}{\barwedge} OB'C', \quad \text{where} \quad S = BB' \cdot CC'.$$

By the fundamental theorem this perspectivity is the same as the given projectivity: the join of two corresponding points always passes through this same point S.

---

* For the direct deduction of this corollary from Poncelet's definition of a projectivity, see Robinson (1940, pp. 28÷31) or Hodge and Pedoe (1947, pp. 218–24).

† von Staudt 1847, p. 51, § 108.

EXERCISE

If the sides of a variables triangle pass through three fixed collinear points, while two vertices run along fixed lines, prove that the third vertex will run along a third fixed line concurrent with the other two. (This Pappus's porism, which was the inspiration for much of Maclaurin's work on loci, beginning in 1722.) (*Hint:* Either use the dual of 4·23 or apply 2·26 to two positions of the variable triangle.)

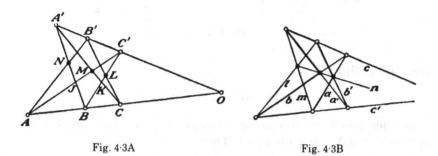

Fig. 4·3A   Fig. 4·3B

4·3 Pappus's theorem. The theorem we are about to prove is especially significant because in some treatments it is taken as an axiom, instead of 3·51. The resulting geometry is more general, as it can be developed without any appeal to continuity.

4·31 Pappus's theorem. *If alternate vertices of a hexagon lie on two lines, the three pairs of opposite sides meet in three collinear points.*

Proof:[*] Let $AB'CA'BC'$ be the hexagon, so that the points to be proved collinear are

$$L = BC' \cdot CB', \qquad M = CA' \cdot AC', \qquad N = AB' \cdot BA',$$

as in Fig. 4·3A. Using further points

$$J = AC' \cdot BA', \qquad K = BC' \cdot CA', \qquad O = AB \cdot A'B',$$

we have

$$A'NJB \overset{A}{\barwedge} A'B'C'O \overset{C}{\barwedge} KLC'B.$$

Thus B is an invariant point of the projectivity $A'NJ \barwedge KLC'$. By 4·23 this is a perspectivity, namely, $A'NJ \overset{M}{\barwedge} KLC'$ (since the joins $A'K$ and JC' pass through M). Hence NL passes through M.

Pappus's theorem suggests a more symmetrical construction (Fig. 4·3C) to replace Fig. 2·7D. Given four points A, B, C, X on one line and three points A', B', C' on another, we can locate X' such that $ABCX \barwedge A'B'C'X'$ as the point $A'B' \cdot AF$, where F is $XA' \cdot o$, and o is

---

* O'Hara and Ward 1937, p. 53.

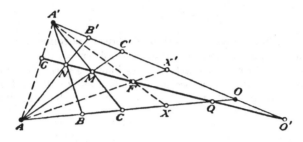

Fig. 4·3C

the 'Pappus line'

$$(CA' \cdot AC')(AB' \cdot BA'),$$

sometimes called the *axis* of the projectivity. To see this, let G be the point where AA' meets the axis o. Then

$$ABCX \overset{A'}{\overline{\wedge}} GNMF \overset{A}{\overline{\wedge}} A'B'C'X'.$$

It is remarkable that Pappus's theorem, when used as an axiom, can take the place of Axiom 2·25. In fact, *Desargues's theorem can be deduced from* 2·21–2·24 *and* 4·31. The following is a simplified version of Hessenberg's proof:[*]

Using the notation of Fig. 1·5B or the frontispiece, name the four extra points

$$S = PR \cdot Q'R', \qquad T = PQ' \cdot RR', \qquad U = PQ \cdot OS, \qquad V = P'Q' \cdot OS.$$

Then the Pappus hexagon $QROSQ'P$ makes A, T, U collinear,
$$P'R'OSPQ' \text{ makes } B, T, V \text{ collinear,}$$
and $\qquad\qquad\qquad Q'SPUTV$ makes A, B, C collinear.

EXERCISES

1. Show the Pappus's theorem is its own dual (see Fig. 4·3B).

2. Let the points A, B, C, A', B', C', L, M, N of Fig. 4·3A be renamed A_1, B_1, C_1, A_2, B_2, C_2, A_3, B_3, C_3. Observe that A_i, B_j, C_k are collinear whenever $i + j + k$ is a multiple of 3.[†]

3. Given a triangle $A_1A_2A_3$ and two points B_1, B_2, locate a point B_3 such that the lines A_1B_1, A_2B_3, A_3B_2 are concurrent, while also A_1B_3, A_2B_2, A_3B_1 are concurrent. Prove that then the lines A_1B_2, A_2B_1, A_3B_3 are concurrent. (In other words, if two triangles are doubly perspective, they are triply perspective.)[‡]

---

[*] Pasch and Dehn 1926, p. 227; Baker 1943, pp. 25–6; Hodge and Pedoe 1947, p. 272.

[†] Levi 1929, p. 108.

[‡] Veblen and Young 1910, p. 100.

4. Show that the Pappus configuration of nine points and nine lines may be regarded (in six ways) as consisting of a cycle of three triangles, such that the three sides of each pass through the three vertices of the next.* (*Hint:* Let one of the triangles be ABN (or $A_1 B_1 C_3$).)

5. Let the axis of $ABC \barwedge A'B'C'$ (Fig. 4·3C) meet AB in Q and $A'B$ in O'. Show that $ABCOQ \barwedge A'B'C'O'O$.

4·4 Classification of projectivities. In this and the next two sections, 'projectivity' will mean 'projectivity between ranges on one line'. Such a projectivity may be either *direct* (sense-preserving) or *opposite* (sense-reversing). The identity is, of course, direct; by 4·12, no other projectivity can have more than two invariant points.

A projectivity having no invariant point is said to be *elliptic*. A projectivity having one invariant point is said to be *parabolic*. A projectivity having two invariant points is said to be *hyperbolic*.

(These names will be justified when we come to consider affine geometry, where the various kinds of conic have 0, 1, or 2 points at infinity.)

By 3·61, every opposite projectivity is hyperbolic; therefore every elliptic or parabolic projectivity is direct. Thus the two methods of classification are related as in the following:

Table of projectivities on one line

Direct			Opposite
The identity (∞)	Elliptic (0)	Parabolic (1)	Hyperbolic (2)

(The numbers of invariant points are given in parentheses.)

This shows that, apart from the identity, there are four possible kinds of projectivity (for superposed ranges): elliptic, parabolic, direct hyperbolic, and opposite. We proceed to prove that all four kinds actually exist.

Special hyperbolic projectivities of the two kinds have already appeared in 3·43 and 3·41.

An instance of an elliptic projectivity is afforded by

$$ABC \barwedge BCA,$$

where A, B, C are any three collinear points. These points themselves are obviously not invariant, and each of the three segments BC/A, CA/B, AB/C is related to another one; hence there is no place for an invariant point anywhere. The actual construction for an elliptic projectivity requires the full allowance of *three* perspectivities. For if a

---

* Hessenberg 1930, p. 69.

projectivity $X \barwedge X'$ on one line is the product of two perspectivities

$$X \overline{\wedge} X_0 \overline{\wedge} X',$$

there must be an invariant point where the line of X's meets the line of X_0's. Conversely,

4·41 *Every parabolic or hyperbolic projectivity (with a given invariant point) can be constructed as the product of two perspectivities.*

Proof: By the fundamental theorem 4·21, a projectivity having an invariant point M is uniquely determined by the relation

$$MAB \barwedge MA'B'.$$

Choose any two points A_0 and B_0 collinear with M, and construct

$$R = AA_0 \cdot BB_0, \qquad S = A_0A' \cdot B_0B',$$

as in Fig. 4·4A. Then we can locate the X' for a given X by means of the two perspectivities

$$MABX \overset{R}{\overline{\wedge}} MA_0B_0X_0 \overset{S}{\overline{\wedge}} MA'B'X'.$$

Here M is the given invariant point. Any other invariant point N must project from R and S into the same point N_0 on A_0B_0; hence it must lie on RS. Thus the projectivity is parabolic if RS passes through M, and hyperbolic otherwise.

By 4·21, a hyperbolic projectivity is determined when both invariant points and one pair of corresponding points are given, particularly when $MNA \barwedge MNA'$.

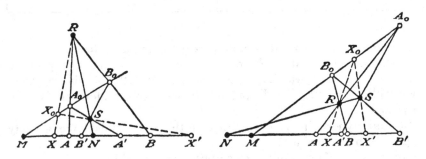

Fig. 4·4A

Such a projectivity exists for any four collinear points M, N, A, A'. To construct it, choose any two points R and S collinear with N, locate $A_0 = AR \cdot A'S$, and use the line MA_0 as before. For any X on MA we have

$$MNAX \overset{R}{\overline{\wedge}} MN_0A_0X_0 \overset{S}{\overline{\wedge}} MNA'X'$$

as in Fig. 4·4B.

The pair AA' may or may not separate the pair MN. In the latter case, by 3·22, S(MNA) = S(MNA'). Hence:

4·42   *The hyperbolic projectivity $MNA \barwedge MNA'$ is opposite if $MN||AA'$ and direct otherwise.*[*]

The above construction remains valid when N coincides with M (so that the projectivity is no longer hyperbolic but parabolic, as in Fig. 4·4C). In this case, if X is taken as A', the quadrangle RSA_0X_0 gives H(MA', AX'). Conversely, such a figure can be reconstructed from any harmonic set MA', AA'', and then M is the only invariant point of the projectivity $MAA' \barwedge MA'A''$. Hence:

4·43   *The projectivity $MAA' \barwedge MA'A''$ is parabolic if*

$$H(MA', AA'')$$

and hyperbolic otherwise.

4·44   Corollary. *A parabolic projectivity is determined when its invariant point and one pair of corresponding points are given.*

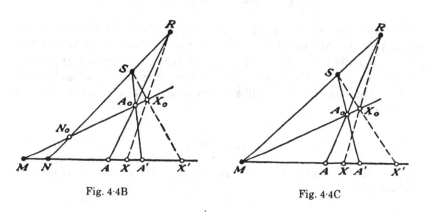

Fig. 4·4B   Fig. 4·4C

Such a projectivity is naturally denoted by

$$MMA \barwedge MMA';$$

therefore we may say that the relation H(MA', AA'') is equivalent to $MMAA' \barwedge MMA'A''$.

This notation is justified by its transitivity:

4·45   *The product of two parabolic projectivities having the same invariant point is another such parabolic projectivity (if it is not merely the identity).*

---

[*] Enriques 1930, p. 101.

Proof: Clearly, the common invariant point of the two given projectivities is still invariant for the product. If any other point A were invariant too, the first projectivity would take A to some different point A' and the second would take A' back to A. By 4·44 the second would then be just the inverse of the first. Hence, apart from that trivial case, the product is parabolic with the same invariant point.

EXERCISES

1. Which part of Fig. 4·4A is *direct*?

2. Draw a figure to illustrate $MMAA'A'' \barwedge MMA'A''A'''$.

3. Let X be a variable point collinear with A, B, C, D, and suppose H(AB, XY), H(CD, $X'Y$). Prove that the projectivity $X \barwedge X'$ is elliptic if $AB//CD$, parabolic if $D = B$.

4·5 Periodic projectivities.

Suppose a given projectivity relates X to X', X' to X'', and so on:

$$XX'X'' \cdots X^{(n-1)} \barwedge X'X''X''' \cdots X^{(n)}.$$

If $X^{(n)}$ coincides with X for three (and therefore all) positions of X, the projectivity is said to be *periodic* and the smallest n for which this happens is called the *period*. Thus the identity is of period 1, the correspondence between harmonic conjugates with respect to M and N (see 3·41) is of period 2, and the elliptic projectivity $ABC \barwedge BCA$ (p. 45) is of period 3.

EXERCISES

1. Show that, if H(AC, BD), $ABC \barwedge BCD$ is an elliptic projectivity of period 4. (*Hint: ABCD $\barwedge$ BCDA.*)

2. Show that a parabolic projectivity cannot be periodic. (*Hint:*

$$\text{S}(MXX') = \text{S}(MXX'') = \text{S}(MXX''') = \cdots = \text{S}(MXX^{(n)}), \text{ so } (X^{(n)} \neq X.)$$

3. Prove that every periodic hyperbolic projectivity is opposite.

4. Prove that the only possible period for a hyperbolic projectivity is 2. (Otherwise the 'squared' projectivity $X \barwedge X''$ would be periodic, hyperbolic, and direct.)

4·6 Involution.

Desargues defined an involution as the relation between pairs of points on a line whose distances from a fixed point have a constant product (positive or negative). The following projective definition is due to von Staudt.*

---

* 1847, pp. 119–20.

An *involution* is a projectivity of period 2:

$$XX' \barwedge X'X.$$

It is remarkable that this relation holds for all positions of X if it holds for any one position; in other words,

4·61 *A projectivity that interchanges two points is necessarily an involution.*

Proof: Suppose we are given $AA' \barwedge A'A$. Consider any point X, and suppose $X \barwedge X'$. By the fundamental theorem 4·21, the given projectivity is the only one in which

$$AA'X \barwedge A'AX'.$$

By 2·71 there is a projectivity in which $AA'XX' \barwedge A'AX'X$. Hence this is the same as the given projectivity, and XX' is a doubly corresponding pair. Since X is quite arbitrary, this proves that the projectivity is an involution.

4·62 Corollary. *An involution is determined by any two of its pairs.*
Notation: The involution $AA'BB' \barwedge A'AB'B$ is denoted by

$$(AA')(BB').$$

Either pair may be replaced by an invariant point repeated: the involution $AA'M \barwedge A'AM$ is denoted by $(AA')(MM)$.

4·63 *If an involution has one invariant point, it has another, and the involution is just the correspondence between harmonic conjugates with respect to these two points.*
Proof: Consider the involution $(AA')(MM)$, and let N be the harmonic conjugate of M with respect to A and A'. Then N is also the harmonic conjugate of M with respect to A' and A; but the involution, being a projectivity, preserves the harmonic relation. Hence N is a second invariant point (distinct from M, by 3·33). If another pair XX' is used instead of AA', we still obtain the same harmonic conjugate N, since otherwise the involution would have three invariant points.
Corollary. *There is no parabolic involution.**

Since the hyperbolic involution $AA'MN \barwedge A'AMN$ is completely determined by its two invariant points, we may denote it by

$$(MM)(NN).$$

From 3·41 we immediately deduce the following:

---

* For some purposes it is convenient to admit the 'degenerate involution' that relates every point X to one fixed point M. The appropriate symbol is $(AM)(BM)$, where $A \neq B$.

4·64   *Every hyperbolic involution is opposite.*

Thus every direct involution is elliptic, and for involutions the table on p. 45 becomes simply

Direct	Opposite
Elliptic (0)	Hyperbolic (2)

Any four collinear points A, A', B, B' determine an involution $(AA')(BB')$ or $AA'B \; \overline{\wedge} \; A'AB'$. If the involution is hyperbolic, the two pairs cannot separate each other (see 3·42); but if the involution is elliptic, it is direct,

$$S(AA'B) = S(A'AB') \neq S(AA'B'),$$

and $AA'//BB'$. Hence:

4·65   *The involution $(AA')(BB')$ is elliptic if $AA'//BB'$ and hyperbolic otherwise.*

We now recognize the points M and N of 3·62 as the invariant points of the hyperbolic involution $(AB)(CD)$.

The following criterion is often useful:

4·66   *A necessary and sufficient condition for three pairs AA', BB', CC' to belong to an involution is $ABCC' \; \overline{\wedge} \; B'A'CC'$.*

Proof: If CC' is a pair of $(AA')(BB')$, we have, by 2·71,

$$ABCC' \; \overline{\wedge} \; A'B'C'C \; \overline{\wedge} \; B'A'CC'.$$

Conversely, the relation $ABCC' \; \overline{\wedge} \; B'A'CC' \; \overline{\wedge} \; A'B'C'C$ implies that the three pairs belong to an involution. (We may have $A = A'$ or $B = B'$, but the nature of the proof requires $C \neq C'$.)

Changing the notation, we may say that a necessary and sufficient condition for the pair MN to belong to the involution $(AB')(BA')$ is

$$MNAB \; \overline{\wedge} \; MNA'B'.$$

Of course, $(AB')(BA')$ is the same as $(AB')(A'B)$; hence:

4·67   *The relation $MNAB \; \overline{\wedge} \; MNA'B'$ is equivalent* to*

$$MNAA' \; \overline{\wedge} \; MNBB'.$$

---

* von Staudt 1847, p. 59, §120. In this theorem it is the two *relations* that are equivalent (each implying the other); the two *projectivities* are, of course, distinct.

If two involutions $(AA_1)(BB_1)$ and $(A'A_1)(B'B_1)$ have a common pair MN, the above remarks show that

$$MNAB \barwedge MNB_1A_1 \barwedge MNA'B'.$$

Hence:

4·68 *If MN is a pair of each of the involutions $(AA_1)(BB_1)$ and $(A'A_1)(B'B_1)$, it is also a pair of $(AB')(BA')$.*

One reason for the importance of involutions is apparent in the following theorem:

4·69 *Any one-dimensional projectivity may be expressed as the product of two involutions.**

Proof: Let the given projectivity transform any noninvariant point A into A' and A' into A''. Then its product with the involution $(AA'')(A'A')$ transforms the pair AA' into $A'A$. Hence the product is itself an involution, and the given projectivity is the product of these two involutions, since the 'square' of an involution is the identity. (In symbols, if the given projectivity is T, the first involution I, and the second J, we have $J = TI$, whence $JI = TI^2 = T$.)

EXERCISES

1. Show that the relation $MNAB \barwedge MNBA$ is equivalent to

$$\mathrm{H}(AB, MN).\dagger$$

2. If a hyperbolic projectivity has a pair of corresponding points that are harmonic conjugates with respect to the two invariant points, show that it must be an involution.

3. Given $\mathrm{H}(AA', MN)$ and $\mathrm{H}(BB', MN)$, prove that $A'B'$ is a pair of the involution $(AB)(MN)$.

4. Prove that two involutions, one or both elliptic, on the same line, always have a common pair of corresponding points. (*Hint:* Consider the product $X \barwedge X^{IJ}$ of the two involutions (XX^I) and (XX^J). If one involution is elliptic and the other hyperbolic, the product is hyperbolic by 3·61. Let M be one of the invariant points of this projectivity; then $M = M^{IJ}$ and $M^J = M^I$. On the other hand, if both involutions are elliptic, let A be a particular position of the variable point X. Observe that the product $X \barwedge X^{IJ}$ relates the interval $\overline{AA^I}/A^{JI}$ to the interior interval $\overline{A^{IJ}/A^J}/A$; then use Axiom 3·51.)

5. If the harmonic relations $\mathrm{H}(BC, AA')$, $\mathrm{H}(CA, BB')$, $\mathrm{H}(AB, CC')$ all hold,

---

* Veblen and Young 1910, p. 224.
† von Staudt 1847, p. 58, §118.

prove that the pairs AA', BB', CC' belong to an involution.* (*Hint:* Apply 4·63 to the involution $BCAA' \barwedge ACBB'$, and deduce $ABCC' \barwedge A'B'C'C$.)

6. Prove the converse of §4·5, Ex. 1: If $ABC \barwedge BCD$ is a projectivity of period 4, it is elliptic and H(AC, BD).

4·7 Quadrangular set of six points. When we say that four collinear points A, A', B, B' determine an involution $(AA')(BB')$ we mean that, for any given point C on the line, we can find a companion C' such that CC' is a pair of the involution. For this purpose we may use the same perspectivities as in the proof of 2·71. (Fig. 4·7A is the same as Fig. 2·7C with an arbitrary point C taken on AB.) We have

$$AA'BB'C \stackrel{P}{\barwedge} USTB'Q \stackrel{A}{\barwedge} PVTBR \stackrel{S}{\barwedge} A'AB'BC'.$$

Actually, the points T, U, V are no longer needed. Given A, A', B, B', C, we can construct C' by taking any two points collinear with A', say P and S (Fig. 4·7B), and deducing $Q = CP \cdot B'S$, $R = AQ \cdot BP$. Then

$$C \stackrel{P}{\barwedge} Q \stackrel{A}{\barwedge} R \stackrel{S}{\barwedge} C'.$$

The construction for a harmonic conjugate (Fig. 2·5A) arises as the special case when the line of section joins two diagonal points of the quadrangle, i.e. when $A = A'$ and $B = B'$. We see now that an elliptic involution is just as easy to construct as a hyperbolic involution with given invariant points.

Since the six points A, A', B, B', C, C' lie on the respective sides QR, PS, RP, QS, PQ, RS of the quadrangle $PQRS$, we immediately deduce the following important theorem (due to Pappus):

4·71 *The three pairs of opposite sides of a quadrangle meet any line (not through a vertex†) in three pairs of an involution.*

We naturally call this the *quadrangular involution* determined on the given line by the quadrangle $PQRS$, and we call $AA'BB'CC'$ a *quadrangular set* of six points.

In Fig. 4·4A we used a quadrangle A_0B_0RS to construct the second invariant point N of a hyperbolic projectivity

$$MAB \barwedge MA'B'.$$

The pairs of opposite sides of that quadrangle meet the line AB in the point pairs MN, AB', BA'. Thus 4·66 and 4·67 remain valid when N coincides with M (or C' with C), i.e. when the projectivity is parabolic instead of hyperbolic. Taking 4·45 into consideration, we see that 4·68

* Mathews 1914, p. 88.
† von Staudt 1847, p. 122, § 222. If the given line did pass through a vertex, say S, we should have the 'degenerate involution' $(AS)(BS)$.

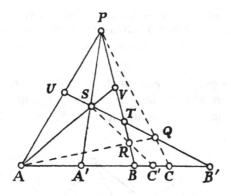

Fig. 4·7A

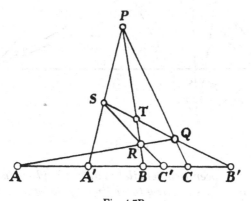

Fig. 4·7B

likewise remains valid when M and N coincide. In other words:

4·72   *M is an invariant point of the involution $(AB')(BA')$ if and only if $MMAB \stackrel{\wedge}{=} MMA'B'$, in which case we have also*

$$MMAA' \stackrel{\wedge}{=} MMBB'.$$

4·73   *If M is an invariant point of each of the involutions*

$$(AA_1)(BB_1) \quad \text{and} \quad (A'A_1)(B'B_1),$$

it is also an invariant point of $(AB')(BA')$.

EXERCISES

1. Take five collinear points A, A', B, B', C, and construct the sixth point of the quadrangular set, as in Fig. 4·7B. Then (below the line, as in Fig. 2·5C) make the analogous construction using C' instead of C. Observe that the new line RS passes through C.

2. Deduce 4·43 from 4·72.

3. How many parabolic projectivities can be found to relate two given points A and B to two given points A' and B'? (None if $AB'//BA'$; two otherwise. See 4·72 and 4·65.)

4·8 Projective pencils. For simplicity we have considered projective ranges; but all our results can be dualized to give properties of projective *pencils*. For instance, the fundamental theorem 4·21 dualizes as follows:

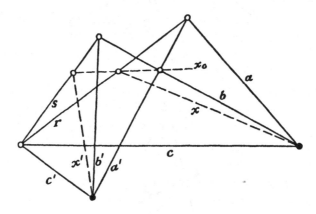

Fig. 4·8A

A projectivity (between two pencils) is determined when three lines of one pencil and the corresponding three lines of the other are given.

When the two pencils have distinct centres, we can dualize Fig. 2·7D to obtain the following construction for the line x' related to a given line x in a given projectivity

$$abc \barwedge a'b'c'.$$

Draw, as in Fig. 4·8A, the lines

$$r = (a \cdot a')(c \cdot c'), \qquad s = (b \cdot b')(c \cdot c'),$$

$$x_0 = (r \cdot x)(b \cdot a'), \qquad x' = (s \cdot x_0)(a' \cdot b').$$

Again, the dual of 4·23 is as follows:
If a projectivity between pencils with distinct centres has an invariant line, it is merely a perspectivity (in the sense of Fig. 2·7B).

EXERCISES

1. Given five concurrent lines m, a, b, a', b', construct the second invariant line of the hyperbolic projectivity $mab \barwedge ma'b'$. (*Hint:* Dualize Fig. 4·4A.)

2. Dualize 4·71. Hence construct the companion of a given line c in a given involution $(aa')(bb')$.

CHAPTER 5

Two-Dimensional Projectivities

We shall find that the one-dimensional projectivity considered in Chapter 4 has two different analogues in two dimensions: one relating points to points and lines to lines, the other relating points to lines and lines to points. The former kind is a collineation, the latter a correlation. Although the general theory is due to von Staudt,* and the names *collineation* and *correlation* to Möbius (1827), some special collineations were used much earlier, e.g. by Newton and La Hire.† Moreover, the classical transformations of the Euclidean plane, viz. translations, rotations, reflexions, and dilatations, all provide instances of collineations. Poncelet considered the relation between the central projections of a plane figure onto another plane from two different centres. He called this special collineation a *homology*. In §5·2 we shall give a purely two-dimensional account of it. Poncelet also considered a special correlation: the polarity induced by a conic. In §5·5, following von Staudt again, we obtain the same transformation without using a conic. We then find that several famous properties of conics are really properties of polarities (which are simply correlations of period two).

5·1 Collineation.‡ We recall that a collineation is a point-to-point correspondence preserving collinearity and consequently preserving the harmonic relation. Thus a collineation induces a projectivity between ranges on corresponding lines and a projectivity between pencils through corresponding points.

5·11 *If the sides of a quadrilateral or the vertices of a quadrangle are invariant, the collineation can only be the identity.*
Proof: Suppose the sides of a given quadrilateral are four invariant lines, then the vertices (where the sides meet in pairs) are six invariant

---
* 1847, pp. 60–66, 125–136.
† See Coolidge 1945, p. 47.
‡ von Staudt 1847, pp. 61, 66, §§123, 130; Cremona 1960, p. 78; Enriques 1930, p. 159.

55

points, three on each side. Hence, by 4·12, every point on each side is invariant. Any other line contains invariant points where it meets the sides and is consequently invariant. Thus the collineation must be the identity. The dual argument gives the same result for a quadrangle.

The fundamental theorem has the following two-dimensional analogue:

5·12 *A collineation is determined when two corresponding quadrilaterals (or quadrangles) are given.*

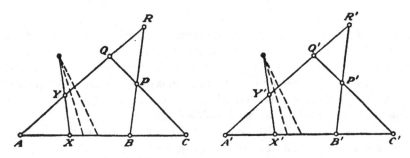

Fig. 5·1A

Proof: Let $ABCPQR$ and $A'B'C'P'Q'R'$ be the two given quadrilaterals. A line of general position may be described as XY, with X on AB and Y on AQ, as in Fig. 5·1A. This determines a line $X'Y'$, where

$$ABCX \barwedge A'B'C'X' \quad \text{and} \quad AQRY \barwedge A'Q'R'Y'.$$

To prove that the correspondence $XY \to X'Y'$ is a collineation, we have to verify that concurrent lines correspond to concurrent lines, i.e. that a pencil of lines XY leads to a pencil of lines $X'Y'$. (It will then follow that collinear points correspond to collinear points.)

For this purpose, let XY vary in a pencil, so that $X \barwedge Y$. By our definition of $X'Y'$ we now have

$$X' \barwedge X \barwedge Y \barwedge Y'.$$

Since A is the invariant point of the perspectivity $X \barwedge Y$, A' must be an invariant point of the projectivity $X' \barwedge Y'$. Hence, by 4·23, this projectivity is again a perspectivity, and $X'Y'$ varies in a pencil, as desired.

Finally, the collineation $ABCPQR \to A'B'C'P'Q'R'$ is unique, by 5·11. In fact, by combining one such collineation with the inverse of another, we should obtain a collineation leaving the quadrilateral invariant.

EXERCISE

Give two reasons why inversion with respect to a circle is *not* a collineation.

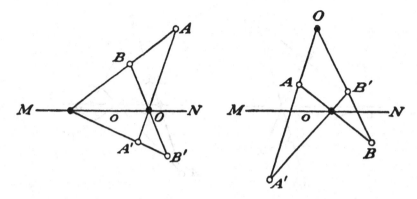

Fig. 5·2A

5·2 Perspective collineation. In particular, a collineation having two invariant points M and N may be described as relating a quadrangle $MNAB$ to a quadrangle $MNA'B'$. It may happen that the two corresponding points $MN \cdot AB$ and $MN \cdot A'B'$ coincide, as in Fig. 5·2A. Then the line $o = MN$ contains three invariant points and consequently consists entirely of invariant points. Thus AA' and BB', meeting o in invariant points, are invariant lines and intersect in an invariant point O.

Every line through O is invariant. For if O does not lie on o, such a line joins O to an invariant point on o; and if O does lie on o, we have three invariant lines through it, namely, o, AA', BB'. (This O is unique; for *any* point X, being joined to two such points by invariant lines, would be invariant, and the collineation would be the identity.) Such a *perspective* collineation, leaving invariant every line through a certain point O and every point on a certain line o, is called an *elation* or a *homology* according as the centre O and axis o are or are not incident.

The above remarks show that every collineation that has three collinear invariant points (or three concurrent invariant lines) is either an elation or a homology.

5·21 *A homology is determined when its centre and axis and one pair of corresponding points (collinear with the centre) are given.**

Proof: Let AA' be the given pair, collinear with the centre O. Any point X (not on OA) determines $C = AX \cdot o$ and $X' = OX \cdot CA'$, as in Fig.

---

* Veblen and Young 1910, p. 72. This use of the word *elation* is apparently due to Lie (1893, p. 262). Poncelet (1865, pp. 155–69) called every perspective collineation a homology, and Enriques (1930, p. 163) distinguished the elation as a 'special' homology.

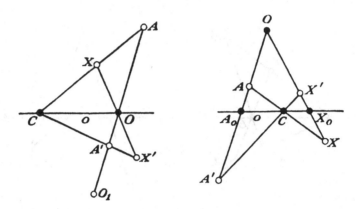

Fig. 5·2B

5·2B. Since all points on o and all lines through O are invariant, the collineation must relate $X = OX \cdot CA$ to the point X' so defined. Similarly

5·22 *An elation is determined when its axis and one pair of corresponding points are given.*

The relation between corresponding points on a given line through O is

$$X \overset{A}{\barwedge} C \overset{A'}{\barwedge} X'.$$

The only possible invariant points of this projectivity $X \barwedge X'$ are on AA' or on o. Hence:

5·23 *An elation or a homology induces a parabolic or hyperbolic projectivity (respectively) on any line through its centre.*

Turning to Fig. 2·2A, we observe that the homology that takes P to P' (with centre O and axis ABC) also takes Q to Q' and R to R'. In the special case when O lies on ABC, we have an elation instead of a homology. Hence:

5·24 *Any pair of Desargues triangles are related by a homology or an elation.*

All the invariant points of an elation lie on its axis. Conversely, a collineation that has a line of invariant points and no others can only be an elation. These remarks will enable us to prove the following:

5·25 *The product of two elations having the same axis is another such elation (if it is not merely the identity).*

Proof: Clearly, each point on the axis is invariant. If any other point A were invariant, too, the first elation would take A to some different point A' and the second would take A' back to A. By 5·22 the second would then be just the inverse of the first. Hence, apart from that trivial case, all the invariant points of the product must lie on the axis (cf. 4·45).

Consider once more the homology determined by O, o, A, and A', as in the second part of Fig. 5·2B. Let OA and OX meet o in A_0 and X_0. The homology is said to be *harmonic* if

$$H(OA_0, AA').$$

In this case we can simply locate X' as the harmonic conjugate of X with respect to O and X_0. Hence:

5·26 *A harmonic homology is determined when its centre and axis are given.*

By an argument similar to that used in proving 5·25, we have the following:

5·27 *The product of two harmonic homologies having the same axis is an elation.*

Conversely:

5·8 *An elation with axis o may be expressed as the product of two harmonic homologies having this same axis o.*

Proof: Let the elation be determined by o, A, and A', as in the first part of Fig. 5·2B, and let O_1 be the harmonic conjugate of $O = AA' \cdot o$ with respect to A and A'. Then the harmonic homologies with centres A and O_1 will have the desired effect, since the first leaves A invariant, while the second takes it to A'.

EXERCISES

1. Show that the central projections of a plane figure onto another plane from two different centres are related by an elation or homology. (*Hint:* The relation is a collineation with a line of invariant points where the two planes intersect.)

2. Justify the above statement that a collineation having a line of invariant points and no others can only be an elation.

5·3 Involutory collineation. A collineation may be *periodic* according to the definition in §4·5. A collineation of period 2 is said to be *involutory*. For instance, a harmonic homology is involutory. Conversely,

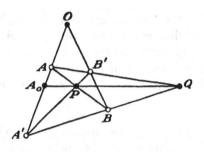

Fig. 5·3A

5·31 *Every involutory collineation is a harmonic homology.*

Proof: Let the given involutory collineation interchange the pair of points AA' and also another pair BB' (not on the line AA'). The invariant lines AA' and BB' intersect in an invariant point O, as in Fig. 5·3A. Since the collineation interchanges the pair of lines AB, $A'B'$ and likewise AB', $A'B$, the two points

$$P = AB \cdot A'B' \quad \text{and} \quad Q = AB' \cdot A'B$$

are invariant. Moreover, the two invariant lines AA' and PQ meet in a third invariant point A_0 on PQ. Hence the collineation is either an elation or a homology. Since the invariant point O does not lie on PQ, it must be a homology; and since

$$\text{H}(AA', OA_0),$$

it is a *harmonic* homology.

5·32 *Two harmonic homologies commute if and only if the centre of each lies on the axis of the other.*

Proof: Let two harmonic homologies H, H' have centres O, O' and axes o, o'. If O lies on o' and O' on o, any two points that are harmonic conjugates with respect to O and O' are interchanged by each homology. Hence the product HH' leaves invariant every point on OO' and, similarly, every line through $o \cdot o'$. Thus HH' is a homology. To see that it is harmonic, we consider two points that are harmonic conjugates with respect to O and $o \cdot o'$. These are interchanged by H but invariant for H'. Thus HH', and similarly $H'H$, is the harmonic homology with centre $o \cdot o'$ and axis OO'.

Conversely, if H and H' commute, their product HH', being equal to its inverse $H'H$, is an involutory collineation, i.e. another harmonic homology. Now, H' transforms any point X on o into a point $X^{H'}$ on $o^{H'}$. Since X is invariant for H, the point

$$X^{H'} = X^{HH'} = X^{H'H}$$

is likewise invariant for H. But $X^{H'}$ may be *any* point on $o^{H'}$. Therefore $o^{H'}$ coincides with o, the axis of H; that is, o is invariant for H' and either coincides with o' or passes through O'. The former possibility is ruled out since, by 5·27, the product HH' would then be an elation. Hence o passes through O', and similarly o' through O.

EXERCISE

Show that the product of three harmonic homologies, whose centres and axes are the vertices and sides of a triangle, is the identity.

5·4 Correlation. We come now to the second kind of two-dimensional projectivity. A *correlation* is a point-to-line correspondence relating collinear points to concurrent lines; it therefore relates concurrent lines to collinear points. Incidences are dualized: we have $X \to x'$ and $x \to X'$, where line x' passes through point X' if and only if point X lies on line x. The range of points X on a given line x corresponds to a pencil of lines x' through the corresponding point X'. Since a quadrangle corresponds to a quadrilateral, four positions of X forming a harmonic set of points correspond to four positions of x' forming a harmonic set of lines. Thus a correlation induces a projectivity between any range and the corresponding pencil.

5·41 *A correlation is determined when a quadrilateral and the corresponding quadrangle are given.*[*]

Proof: Let $ABCPQR$ and $a'b'c'p'q'r'$ be the given quadrilateral and quadrangle, as in Fig. 5·4A. A line XY, with X on AB and Y on AQ, determines a point $x' \cdot y'$, where

$$ABCX \barwedge a'b'c'x' \quad \text{and} \quad AQRY \barwedge a'q'r'y'.$$

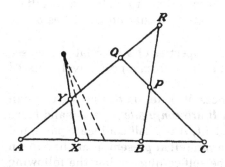 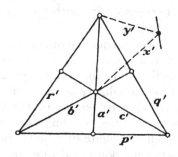

Fig. 5·4A

---

[*] Veblen and Young 1910, p. 264.

To prove that the correspondence $XY \rightarrow x' \cdot y'$ is a correlation, we have to verify that a pencil of lines XY leads to a range of points $x' \cdot y'$.

Let XY vary in a pencil, so that $X \barwedge Y$ and therefore $x' \barwedge y'$. Since A is the invariant point of the perspectivity $X \barwedge Y$, a' must be an invariant line of the projectivity $x' \barwedge y'$. Hence, by the dual of 4·23, this projectivity is a perspectivity, and $x' \cdot y'$ varies in a range as desired.

Since the product of two correlations is a collineation, the uniqueness of the correlation $ABCPQR \rightarrow a'b'c'p'q'r'$ is another consequence of 5·11. In fact, by combining one such correlation with the inverse of another, we should obtain a collineation leaving the quadrilateral invariant.

EXERCISE

Show that any collineation can be expressed as the product of two correlations, one of which may be arbitrarily assigned.

5·5   Polarity. In general, a correlation relates a point X to a line x' and relates this line to a new point X''. The correlation is involutory (i.e. of period 2) if X'' always coincides with X, in which case we may omit the prime ['] without causing any confusion. An involutory correlation is called a *polarity*. Thus a polarity is a correlation that relates X to x, and vice versa. Following Servois and Gergonne, we call X the *pole* of x and x the *polar* of X.

This terminology may be justified as follows: The section of a sphere by a plane through the centre is a great circle, and the 'axis' of this circle meets the sphere in two *poles* (e.g. the North Pole and South Pole are the poles of the equator). When we make a gnomonic projection (from the centre onto an arbitrary plane), the great circle and the two poles yield a straight line and a single point, the *pole* of the line. This is easily seen to satisfy the above description of a polarity. (It differs from the polarity with respect to a conic, in that no point lies on its own polar.)

As a consequence of the general properties of a correlation, we see that the polars of all the points on a line a form a projectively related pencil of lines through the pole A.

Since a polarity dualizes incidences, if A lies on b, a passes through B. In this case we say that A and B are *conjugate points*, a and b are *conjugate lines*. If A and a are incident, A is a self-conjugate point and a a self-conjugate line. So far as we can tell at present, it might happen that every point and line would be self-conjugate; but the following theorem shows that this is not so:*

---

* Enriques 1930, pp. 184, 185.

5·51 *The join of two self-conjugate points cannot be a self-conjugate line.*

Proof: If the join *a* of two self-conjugate points were a self-conjugate line, it would contain its pole *A* and at least one other self-conjugate point, say *B*. The polar of *B*, containing both *A* and *B*, would coincide with *a*: two distinct points would both have the same polar. This is impossible, since a polarity is a one-to-one correspondence between points and lines.

As a further limitation on the occurrence of self-conjugacy, we shall prove the following:*

5·52 *It is impossible for a line to contain more than two self-conjugate points.*

Proof: Let *A* and *B* be two self-conjugate points on *s*, and let *P* be a point on *AS* or *a*, distinct from *A* and *S*. Let the polar *p* meet *b* in *Q* (see Fig. 5·5A). Then $Q = b \cdot p$ is the pole of $BP = q$, which meets *p* in *R*, say. Also $R = p \cdot q$ is the pole of $PQ = r$, which meets *s* in *C*, say. Finally, $C = r \cdot s$ is the pole of $RS = c$, which meets *s* in *D*, the harmonic conjugate of *C* with respect to *A* and *B*. Now, *C* cannot coincide with *A* or *B*; for then *P* would coincide with *A* or *S*. Hence *C*, not lying on *c*, is not self-conjugate.

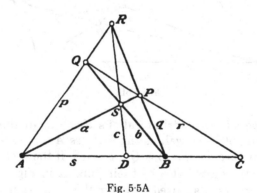

Fig. 5·5A

We thus have, on *s*, two self-conjugate points *A*, *B* and a non-self-conjugate point *C*. But the self-conjugate points on *s* are the invariant points of the projectivity $X \barwedge x \cdot s$ induced on *s* by the given polarity. Hence this projectivity is not the identity, and cannot have more than two invariant points; i.e. the line *s* cannot contain more than two self-conjugate points of the polarity.

We can now easily prove:

5·53 *A polarity induces an involution of conjugate points on any line that is not self-conjugate and an involution of conjugate lines through any point that is not self-conjugate.**

Proof: The projectivity $X \barwedge x \cdot s$ on s relates any non-self-conjugate point C to another point $D = c \cdot s$, whose polar is CS. Hence the same projectivity relates D to C; that is, it *interchanges* C and D. By 4·61, it must be an involution. (In the proof of 5·52 this was a hyperbolic involution; but it can just as well be elliptic.)

Dually, x and XS are paired in the involution of conjugate lines through S.

The following lemma will be needed in the proof of 5·55:

5·54 *If all four sides of a quadrilateral are self-conjugate lines, at most one pair of opposite vertices can be conjugate points.*

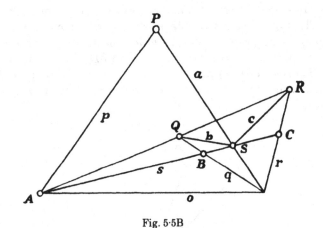

Fig. 5·5B

Proof:† Let *pqrs* be a quadrilateral of self-conjugate lines, so that s contains its own pole S as well as the vertices $A = p \cdot s$, $B = q \cdot s$, $C = r \cdot s$, whose polars are $a = PS$, $b = QS$, $c = RS$. If the vertices $q \cdot r$ and $p \cdot s$ are conjugate points, let o be their join, as in Fig. 5·5B. Then a passes through $q \cdot r$ and is paired with o in the involution of conjugate lines through $q \cdot r$ (see 5·53). Since q and r are the invariant lines of this involution, we have $H(qr, oa)$, and therefore $H(BC, AS)$. Similarly, if $r \cdot p$ were conjugate to $q \cdot s$ we would have $H(CA, BS)$, and if $p \cdot q$ were conjugate to $r \cdot s$ we would have $H(AB, CS)$; but by 3·34 and 3·17, only one of these three harmonic relations can hold.

---

* von Staudt 1847, p. 134, §239. On a self-conjugate line s the conjugate points form a 'degenerate involution'; each point is conjugate to the single point S.

† Kindly supplied by Patrick Du Val.

5·55 Hesse's theorem. *If two pairs of opposite vertices of a quadrilateral are pairs of conjugate points (in a given polarity), then the third pair of opposite vertices is likewise a pair of conjugate points.*

Proof:* Let $APBQCR$ be the quadrilateral, as in Fig. 5·5C, with A conjugate to P, and B to Q. We wish to prove that C is conjugate to R. By 5·54, we may suppose that the quadrilateral has at least one non-self-conjugate side, say $s = ABC$. Thus s does not pass through its pole S. Various special positions of S have to be considered. If S coincides with P, PQ is the polar of B, PB is the polar of C, and R is conjugate to C. The same conclusion follows similarly if S coincides with Q, and still more obviously if S coincides with R. If S lay elsewhere on a side of the triangle PQR, we could deduce absurd coincidences. Hence we may restrict consideration to a 'general' position for S.

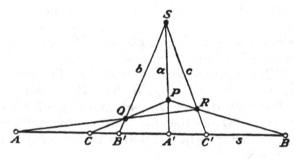

Fig. 5·5C

Let SP, SQ, SR meet s in A', B', C'. Since A lies on s and is conjugate to P, its polar a is PS. Similarly b is QS. Thus the involution of conjugate points on s is $(AA')(BB')$. By 4·71, CC' is another pair of this involution. Hence c is SC'. Since this contains R, C is conjugate to R.

Hesse's theorem dualizes as follows:

5·56 *If two pairs of opposite sides of a quadrangle are pairs of conjugate lines, then the third pair of opposite sides is likewise a pair of conjugate lines.*

EXERCISES

1. Observe that the polarity of 5·52 (Fig. 5·5A) relates the quadrangle $PQRS$ to the quadrilateral $ABCPQR$, and the harmonic set of points A, B, C, D to the harmonic set of lines SA, SB, SD, SC.

2. Justify the statement (in the proof of 5·52) that C coinciding with A or B would make P coincide with A or S.

---

* Suggested by Cremona 1960, p. 238.

5·6 Polar and self-polar triangles.* The *polar triangle* of a given triangle is formed by the polars of the vertices and the poles of the sides.

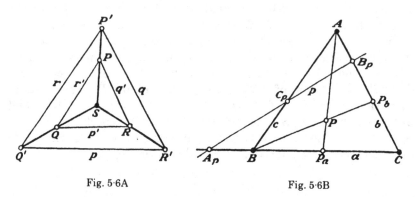

Fig. 5·6A   Fig. 5·6B

5·61 Chasles's theorem. *A triangle and its polar triangle (if distinct) are a pair of Desargues triangles.*

Proof: Let $P'Q'R'$ be the polar triangle of PQR. Consider the quadrangles $PQRS$, where S is $PP' \cdot QQ'$. We have PS (through P') conjugate to $p' = QR$ and QS (through Q') conjugate to $q' = PR$. By 5·56, RS is conjugate to $r' = PQ$ and therefore passes through R', as in Fig. 5·6A. Thus corresponding vertices of the two triangles PQR and $P'Q'R'$ are collinear with S.

A triangle is said to be *self-polar* if each vertex is the pole of the opposite side. Any two points (or lines) that are conjugate but not self-conjugate determine a self-polar triangle; for if A and B are conjugate points on a non-self-conjugate line c, each vertex of triangle ABC is the pole of the opposite side. Any two vertices (or sides) are conjugate. The occurrence of such a triangle is characteristic of a polarity, as the following theorem shows:

5·62 *Any correlation that relates the three vertices of one triangle to the respectively opposite sides is a polarity.*

Proof: Consider the correlation $ABCP \to abcp$, where a, b, c are the sides of the given triangle ABC and P is a point of general position. Let $A_p, B_p, C_p, P_a, P_b, P_c$ denote the respective points

$$a \cdot p, \quad b \cdot p, \quad c \cdot p, \quad a \cdot AP, \quad b \cdot BP, \quad c \cdot CP,$$

as in Fig. 5·6B. The correlation not only relates A, B, C to a, b, c but also relates a, b, c to A, B, C; for it relates $a = BC$ to $A = b \cdot c$, and so on. Moreover, it relates AP to A_p and, consequently, P_a to AA_p. We wish to show that it relates p to P.

---

* von Staudt 1847, pp. 131–5, §§ 234–42; Enriques 1930, p. 182, 187.

Consider the projectivity $X \barwedge x \cdot a$ induced on a. Since $b \cdot a$ and $c \cdot a$ are C and B, this projectivity interchanges B and C; hence it is an involution. Since $P_a A_p$ is a pair of the involution, the correlation works as follows: $A_p \to AP_a$; similarly $B_p \to BP_b$; hence $A_p B_p \to AP_a \cdot BP_b$; that is, $p \to P$.

5·63 Corollary. *A polarity is determined when a self-polar triangle and one further pole and polar are given.*

Notation: The polarity with self-polar triangle ABC, relating P and p, is denoted by

$$(ABC)(Pp).$$

In using this symbol, we assume that P does not lie on a side of triangle ABC and that p does not pass through a vertex.

We proceed to describe a construction (Fig. 5·6C) for the polar of any point X in a given polarity $(ABC)(Pp)$. If X is distinct from P, it cannot lie on more than one of the lines AP, BP, CP, and we may suppose A, B, C to be named in such an order that X does not lie on either of AP, BP. The construction is as follows:

5·64 *The polar of any point X in the polarity $(ABC)(Pp)$ is the line $[AP \cdot (a \cdot PX)(p \cdot AX)][BP \cdot (b \cdot PX)(p \cdot BX)]$.*

Proof: We have to show that the polar of X is the line YZ determined by

$$A_1 = a \cdot PX, \quad E = p \cdot AX, \quad Y = AP \cdot A_1 E,$$

$$B_1 = b \cdot PX, \quad F = p \cdot BX, \quad Z = BP \cdot B_1 F.$$

By Hesse's theorem (our 5·55), applied to the quadrilateral AA_1PEXY (Fig. 5·6C), since AA_1 and PE are pairs of conjugate points, XY is another pair of conjugate points. Thus X is conjugate to Y and similarly to Z. Therefore $x = YZ$.

The above construction fails if X lies on p (since then Y and Z both coincide with P). In this case, take an arbitrary point Q and construct

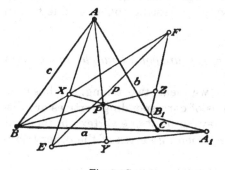

Fig. 5·6C

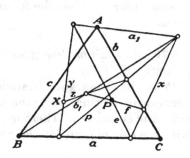

Fig. 5·6D

its polar q. If q happens to pass through the given point X (on p), then the polar of X is PQ, and the problem is solved. If not, we can use the above method to find the polar of X in the polarity $(ABC)(Qq)$.

In 5·63 a polarity was determined by means of a self-polar triangle and one further pole and polar. Another way would be by means of a *self-polar pentagon* such as PBC_pB_pC, that is, a pentagon whose vertices are the poles of the respectively opposite sides. In fact:

5·65 *The correlation that relates four vertices of a pentagon to the respectively opposite sides is a polarity and relates the fifth vertex to the fifth side.*

Proof: The correlation that relates the vertices Q, R, S, T of a pentagon $PQRST$ to the sides ST, TP, PQ, QR also relates the point $A = QR \cdot ST$ to the line $(ST \cdot TP)(PQ \cdot QR) = TQ$. Thus it relates each vertex of triangle AQT to the opposite side and is a polarity, by 5·62. Finally, since it relates RS to $TP \cdot PQ = P$, it also relates P to RS.

EXERCISES

1. Let AA', BB', CC' be a quadrangular set of points and S any point outside their line. Prove that any correlation which relates S, A, B, C to AB, SA', SB', SC' is a polarity (Fig. 5·5C).

2. Show how the construction 5·64 becomes partly indeterminate when P is given on BC or AB (and p through A or C, respectively). *Hint:* If P lies on BC, so also does Z; Y may be any point on AP, and x any line through Z. If P lies on AB, Y and Z coincide; for then ABY is the Pappus line of the hexagon A_1EXFB_1C (see 4·31).

3. By dualizing 5·64, derive X from x (Fig. 5·6D).

4. Consider the self-polar pentagon $PQRST$ of Theorem 5·65. Let an arbitrary line through P meet ST in U and QR in V. Prove that RU and SV are conjugate lines. Deduce a construction for the polar of any given point X.

5·7   The self-polarity of the Desargues configuration.* Chasles's theorem (our 5·61) has the following interesting converse, due to von Staudt:

5·71 *Any pair of Desargues triangles are polar triangles in a certain polarity.*

Proof: By 'Desargues triangles' we mean two triangles whose six vertices are distinct while the joins of corresponding vertices are distinct and concurrent. There is always at least one vertex not incident with the side opposite to the corresponding vertex of the other trian-

---

* This section may be omitted on first reading or in a short course.

gle; for it is clearly impossible for each triangle to be inscribed in the other.

Let PQR and $P'Q'R'$ be two such triangles, with the vertex P of the former not lying on the side $Q'R'$ of the latter. Since Q and R do not both lie on $Q'R'$, we may assume (by interchanging their names if necessary) that Q does not lie on $Q'R'$. Let S be the common point of PP', QQ', RR', as in Fig. 5·7A. The point R', not lying on QQ', cannot coincide with S and, therefore, cannot lie on PP'. Thus the triangle PAK, where $A = QR \cdot Q'R'$ and $K = PS \cdot Q'R'$, has no side through Q and no vertex on $q = R'P'$.

Consider the polarity $(PAK)(Qq)$. Since A is conjugate to S (on PK) while Q is conjugate to R' (on q), Hesse's theorem shows that Q' is conjugate to R. Since A is conjugate to P' (on KP) while P is conjugate to both Q' and R' (on AK), the polars of P', Q', R' are respectively AQ, RP, PQ. Since these are the sides of the triangle PQR, the theorem is proved.

EXERCISES

1. Let Π denote the polarity of 5·71 and Γ the homology or elation of 5·24. Prove that $\Gamma\Pi = (PQR)(Ss)$ and $\Pi\Gamma = (P'Q'R')(Ss)$. Deduce that these two polarities coincide when Γ is a harmonic homology.

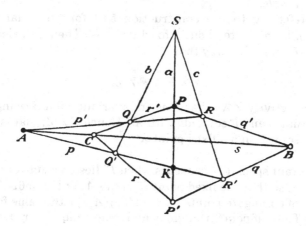

Fig. 5·7A

2. Prove that the polarity $(ABC')(Nn)$ interchanges the nine points and nine lines of the Pappus configuration (Figs. 4·3A, B) if and only if the lines LA', MB', CN are concurrent. Hint: If those three lines are concurrent, consider the polarity of 5·71 as applied to triangles LMC and $A'B'N$. Show that B is the pole of AC', by applying 5·56 to the quadrangle $AC'BB'$.* (Since the general Pappus configuration

---

* This method was devised by Alex Rosenberg when he was an undergraduate at the University of Toronto.

is self-dual without being self-polar, the old controversy between Poncelet and Gergonne is settled in the latter's favour.)

3. Verify that the polarity $(PAK)(Qq)$ of 5·71 transforms all the ten points P, Q, R, P', Q', R', S, A, B, C of the Desargues configuration into the ten lines that contain these points by threes.

4. Show that the Desargues configuration contains 12 self-polar pentagons such as $ABPSQ'$, any one of which could be used to determine von Staudt's polarity (cf. § 2·2, Ex. 4).

5·8   Pencil and range of polarities. The various polarities that have a given self-polar triangle and a given pair of conjugate points (or lines) are said to form a pencil (or range) of polarities. Thus, given a self-polar triangle ABC, the polarities $(ABC)(Pp)$ form a *pencil* if P is fixed while p varies in a pencil of lines; dually, they form a *range* if p is fixed while P varies along a line. Thus any two polarities that have a common self-polar triangle belong to a definite pencil and to a definite range.

5·81   *The polars of a fixed point* X, *with respect to a pencil of polarities, form a pencil of lines, except when* X *is a vertex of the common self-polar triangle.*

Proof: Referring to the construction 5·64 for the polar of X in $(ABC)(Pp)$, let p rotate about a fixed point P'. Then Y varies on AP, and Z on BP, in such a way that

$$Y \overset{A_1}{\barwedge} E \overset{P'}{\barwedge} F \overset{B_1}{\barwedge} Z.$$

But the projectivity $Y \barwedge Z$ has P as an invariant point, arising when p is $P'X$ (so that E and F coincide with X). Hence $Y \overline{\barwedge} Z$; that is, the line $x = YZ$ passes through a fixed point X'.

An important special case occurs when P' lies on a side of the triangle, say on a, so that the fixed point P' is the A_p of Fig. 5·6B. Then the involution of conjugate points on a is $(BC)(A_p P_a)$, the same for all the polarities. The self-polar triangle is no longer unique; for A, with any pair of this involution would form a self-polar triangle that could be used instead of ABC. Any point X projects from A into a point X_a on a, and the polars x form a pencil of lines through A_x, the companion of X_a in the involution $(BC)(A_p P_a)$. Conversely, if any line x meets a in A_x, its poles X all lie on the fixed line AX_a, where X_a is the companion of A_x. Hence this kind of pencil of polarities is at the same time a range. Let us simply call it a *self-dual system* of polarities. To sum up:

5·82   *A self-dual system admits a line* a *on which the involution of conjugate points is the same for all the polarities. The polars of any point*

P form a pencil of lines through a fixed point A_p on a, and the poles of any line p form a range on a fixed line AP_a, where A is the pole of a (for all the polarities).

The product of two polarities (or, indeed, of any two correlations) is a collineation. In particular,

5·83 *If two polarities belong to a self-dual system, their product is a homology.*

Proof: Since the polarities induce the same involution of conjugate points on the line a and the same involution of conjugate lines through the point A, their product leaves invariant every point on a and every line through A.

Conversely,

5·84 *Any homology* can be expressed as the product of two polarities belonging to a self-dual system.*

Proof: The homology with centre A and axis BC, relating P_1 to P_2 (on a line through A), is the product of polarities $(ABC)(P_1p)$ and $(ABC)(P_2p)$, where p is an arbitrary line.

EXERCISES

1. Show that the relation between X and X' in 5·81 is symmetric (i.e. involutory) —that the polars of X' all pass through X. (*Hint:* X and X' are conjugate points.)

2. Show that the correspondence $X \to X'$ is not a collineation. (*Hint:* Take X at various positions on AB.)

3. Show that, in the notation of the proof of 5·81, $x \barwedge p$.

4. Show that the two polarities $(ABC)(P_1p)$ and $(ABC)(P_2p)$ of 5·84 will commute if $H(AP_a, P_1P_2)$.

5. Show that the polarities of §5·6, Exs. 1 and 2, belong to self-dual systems.

6. Show that a range of polarities is determined by a self-polar pentagon $PQRST$, where P varies along a line while the other four vertices remain fixed.

5·9 Degenerate polarities. When a pencil of polarities was specified by $(ABC)(Pp)$ with p through P' and P not on a side of the self-polar triangle ABC, we tacitly excluded three apparently possible positions of p, namely, $P'A$, $P'B$, $P'C$. For the sake of completeness, we should try to interpret the symbol

---

* Veblen and Young (1910, p. 265) prove the more general statement that any collineation can be expressed as the product of two polarities; but their construction breaks down if the collineation is involutory.

$$(ABC')(Pp)$$

when p passes through A although P does not lie on a. The construction 5.64 makes Y coincide with A but yields a definite point Z such that $BB_1 PFXZ$ is a quadrilateral. By the dual of 4·71, the polar $x = AZ$ of an arbitrary point X is the mate of AX in the involution

$$(bc)(pp'),$$

where p' is AP. For each pair of lines in this involution, the degenerate polarity relates every point on either line to the other line. Thus the polarity reduces to an involution among the lines through A. It differs from an ordinary polarity in that a line through A, instead of having a unique pole, has a whole range of poles; but the pole of any other line is just the point A.

Referring to the proof of 5·81, we can now say that, when p is $P'A$, x is $X'A$.

Dually, we find another kind of degenerate polarity $(ABC)(Pp)$ by taking P on a while p does not pass through A. The pole of any line x is the mate of $a \cdot x$ in the involution $(BC)(PP')$, where P' is $a \cdot p$. In this sense we can identify the degenerate polarity with this involution on a.

Thus a general pencil of polarities includes three degenerate polarities of the first kind, while a general range of polarities includes three of the second kind; but a self-dual system includes one of each. For, if the system is expressed as $(ABC)(Pp)$ with P fixed while p passes through a fixed point P' on a, we obtain a degenerate member of the first kind by taking p to be $P'A$; but if it is expressed in the same manner with p fixed while P lies on a fixed line p' through A, we obtain a degenerate member of the second kind by taking P to be $p' \cdot a$.

For the sake of completeness we should add that the self-dual system also includes two kinds of *doubly degenerate* polarity:

$$(ABC)(Pa) \quad \text{and} \quad (ABC)(Ap),$$

where P and p are of general position. The former makes every point have the same polar a while the pole of any line x is $x \cdot a$. The latter makes every line have the same pole A while the polar of any point X is XA.

CHAPTER 6

Conics

6·1 Historical remarks. The study of conic sections (or briefly, conics) is said to have begun in 430 B.C., when the Athenians, suffering from a plague, appealed to the oracle at Delos and were told to double the size of Apollo's cubical altar. Let a denote the edge of the original cube and x that of the enlarged one; then the requirement is

$$x^3 = 2a^3,$$

or $x/a = \sqrt[3]{2}$. Hippocrates reduced the problem to that of finding values of x and y to satisfy the equations

$$\frac{a}{x} = \frac{x}{y} = \frac{y}{2a}.$$

Menaechmus, about 340 B.C., gave two solutions: one using the two parabolas $y^2 = 2ax$, $x^2 = ay$, and the other using the latter parabola along with the rectangular hyperbola $xy = 2a^2$. Without the benefit of algebraic notation, this was surely a marvellous achievement. In fact, it shows that Menaechmus came close to anticipating by 2000 years the analytic geometry of Fermat and Descartes. He presumably obtained these curves as plane sections of a right circular cone. (Hence the name *conic section*.) Purely two-dimensional constructions were soon devised by his successors. According to Zeuthen, it was Euclid who first constructed a conic as the locus of a point whose distance from a fixed point (focus) is proportional to its distance from a fixed line (directrix). The names *ellipse, parabola, hyperbola* are due to Apollonius (262–200 B.C.), who discovered an astonishing number of their properties. He even anticipated Steiner's theorem (our 6·52). Some further results were obtained by Pappus, about A.D. 300, but after that time the whole subject was forgotten for twelve centuries.

In fact, no new contribution of any importance was made until 1522, when Verner of Nuremberg derived certain properties of conics by projecting a circle. For the next three centuries, apart from Kepler, Newton, Maclaurin, and Braikenridge, the subject was developed

73

largely by Frenchmen: Desargues, Pascal, Mydorge, and La Hire, and
then Gergonne, Brianchon, Poncelet, and Chasles. Kepler showed how
a parabola is at once the limiting form of an ellipse and of a hyperbola,
thus paving the way for consideration of the general conic. The name
of Braikenridge (1700–59) is not very familiar, but he shares with
Maclaurin the honour of discovering the first non-metrical construction for a conic.

The more recent developments are dominated by the names of a Swiss
and a German: Steiner (1793–1863) and von Staudt. We shall see, in
§ 6·5, how their two ways of approaching the conic may be reconciled.
The polarity induced by a conic is implicit in some of the work of
Apollonius and was clearly understood by La Hire (1640–1718), but it
was von Staudt who turned the tables by allowing the polarity to define the conic. This standpoint provides the most symmetrical definition for conics and emphasizes their self-dual nature, as we shall see in
§ 6·3.

6·2   Elliptic and hyperbolic polarities.* We recall that an involution is hyperbolic or elliptic according as it does or does not admit an
invariant point, that a hyperbolic involution has not merely one but
two invariant points, and that the involution $(AB)(CD)$ is elliptic or
hyperbolic according as D does or does not lie in the segment AB/C.

Analogously, a polarity is said to be *hyperbolic* or *elliptic* according
as it does or does not admit a self-conjugate point, i.e. a point lying on
its own polar. We shall find that a hyperbolic polarity has not merely
one but infinitely many self-conjugate points, forming a curve (in fact
a conic), and that the polarity $(ABC)(Pp)$ is elliptic or hyperbolic according as P does or does not lie in the triangular region ABC/p.

6·21   *If P lies in ABC/p, the polarity $(ABC)(Pp)$ is elliptic.*

Proof: Let p meet the sides of triangle ABC in A_p, B_p, C_p, as in Fig.
6·2A, and let the three lines AP, BP, CP meet the respective sides in P_a,
P_b, P_c. If P lies in the region ABC/p, then P_a lies in the segment BC/A_p
and the involution $(BC)(A_pP_a)$ is elliptic. This is, however, the involution of conjugate points on the line BC. Similarly, the involutions on
the other sides are elliptic too. If instead of p we take another line x
(not through a vertex), we obtain other pairs of these involutions;
hence the points X_a, X_b, X_c (determined on the sides by the pole X) lie
in the respective segments BC/A_x, CA/B_x, AB/C_x. Thus X lies in the
region ABC/x and cannot lie on x. This means that there are no self-
conjugate points, except possibly on a side of triangle ABC; but a
self-conjugate point on a side would be an invariant point of the invo-

---

* von Staudt 1847, p. 133, § 237; Enriques 1930, pp. 187–91.

lution of conjugate points on that side, whereas we have just seen that such an involution is elliptic. This completes the proof.

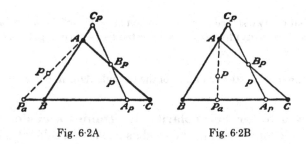

Fig. 6·2A   Fig. 6·2B

6·22 *If P does not lie in ABC/p, but in one of the other three regions, the polarity (ABC)(Pp) is hyperbolic.*

Proof: For definiteness, suppose that P lies in the region penetrated by p through sides a and b, as in Fig. 6·2B. Then A_p and P_a both lie in the same segment BC, and thus the involution $(BC)(A_pP_a)$ is hyperbolic. Since this is the involution of conjugate points on a, its invariant points are self-conjugate.

Similarly, there are two self-conjugate points on b, though none on c.

6·23 Corollary. *Both elliptic and hyperbolic polarities exist.*

The following temporary definitions will be found helpful in § 6·3 but thereafter will be superseded: Let a point that is not self-conjugate be called an E point or an H point according as the involution of conjugate lines through it is elliptic or hyperbolic, and let a line that is not self-conjugate be called an e line or an h line according to the nature of the involution of conjugate points on it. We see at once that in an *elliptic* polarity every point is an E point and every line is an e line; and it emerges from the above proof of 6·22 that any self-polar triangle for a *hyperbolic* polarity has two h sides and one e side and consequently two H vertices and one E vertex. (Of course, the pole of an e line or h line is an E point or H point, respectively.) Self-polar triangles in the two cases are represented diagrammatically in Fig. 6·2C.

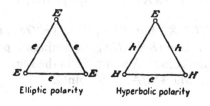

Elliptic polarity   Hyperbolic polarity

Fig. 6·2C

EXERCISES

1. Carry out the construction 5·64 in the special case when P lies on p. Observe that X lies outside the region ABC/x.

2. Show that the two self-conjugate points on an h side (of the self-polar triangle for a hyperbolic polarity) lie respectively on the two self-conjugate lines through the opposite H vertex.

3. Which are the E vertex and e side of the triangle ABC in Fig. 5·6B?

6·3 How a hyperbolic polarity determines a conic. Any line that is not self-conjugate may be used as a side of a self-polar triangle. Thus, in the case of a hyperbolic polarity, every point on an e line is an H point, but an h line contains points of both types as well as two self-conjugate points (which are the invariant points of the involution of conjugate points on the h line). Dually, the pencil of lines through an E point consists entirely of h lines, but the pencil of lines through an H point contains both types as well as two self-conjugate lines. Thus we have through an E point a pencil of h lines, each containing two self-conjugate points, and on an e line a range of H points, through each of which two self-conjugate lines can be drawn. Hence:

6·31 *A hyperbolic polarity admits an infinity of self-conjugate points and an infinity of self-conjugate lines.*

We proceed to prove that the E points and H points on an h line are separated by the two self-conjugate points; that is:

6·32 *The two self-conjugate points on an h line decompose it into two segments, one consisting of E points and the other of H points.*

*Proof:** On an h line, o, let Q and R be the self-conjugate points, and A an H point (see Fig. 6·3A). Let P be one of the self-conjugate points on a, the polar of A, and let a meet o in A_1. Take any point X of the segment QR/A. Let its polar, x, meet a in O, o in X_1, PA in K, and PX in K'. Since $H(QR, AA_1)$ and $H(QR, XX_1)$, one of the two segments QR contains A and X_1 while the other contains A_1 and X. By 3·41, the order of points on the line is either

$$QAX_1RXA_1Q \quad \text{or} \quad QX_1ARA_1XQ.$$

It follows that $AX//A_1X_1$; but $AXA_1X_1 \overset{P}{\barwedge} KK'OX_1$. Hence $KK'//OX_1$. Therefore the involution $(KK')(OX_1)$ of conjugate points on x is elliptic; x is an e line, and X is an E point. Thus the segment QR/A consists entirely of E points, and QR/A_1 of H points.

---

* Kindly supplied by J.A. Jenkins.

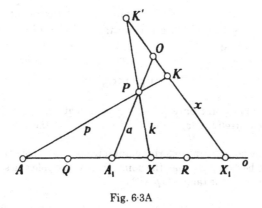

Fig. 6·3A

Dually, the *e* lines and *h* lines through an *H* point are separated by the two self-conjugate lines.

Following von Staudt, we define a *conic* to be the locus of self-conjugate points in a hyperbolic polarity.* An *h* line (meeting the conic in two points, as in Fig. 6·3B) is a *secant*. A self-conjugate line (meeting the conic only at its pole) is a *tangent*, and the pole is its *point of contact*. An *e* line (containing no self-conjugate points) in an *exterior line*. The pole of a secant is an *exterior point* (*H* point), which lies on two tangents (the invariant lines of the involution of conjugate lines through it). Finally, the pole of an exterior line is an *interior point* (*E* point), which is characterized by the fact that no tangents can be drawn through it.

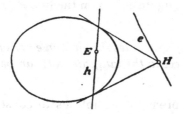

Fig. 6·3B

Thus a conic is essentially a self-dual figure: it is the locus of self-conjugate points and also the envelope of self-conjugate lines. Any of its properties can immediately be dualized by applying the polarity that defines it.

By our remarks about *H* points and *e* lins, every point on an exterior line is an exterior point. Dually, every line through an interior point is

---

* von Staudt 1847, p. 137, § 246; Enriques 1930, pp. 199–201, 261–2.

a secant. Again, any point on a tangent, except the point of contact, is exterior; and every line through a point on the conic, except the tangent there, is a secant.

EXERCISES

1. Show that every point on a tangent is conjugate to the point of contact. Dually, the tangent itself is conjugate to any line through the point of contact.

2. Show that the polar of any exterior point joins the points of contact of the two tangents that can be drawn through the point. Dually, the pole of a secant PQ is the point of intersection of the tangents p and q.

3. If PQR is a triangle inscribed in a conic, the tangents at P, Q, R form a triangle circumscribed about the conic. Prove that these are a pair of Desargues triangles. (*Hint:* Use 5·61.)

4. If the tangents to a given conic meet a second conic in pairs of points, show that the tangents to the second conic at these pairs of points meet on a third conic. (*Hint:* Any construction for the first conic will be transformed, by the polarity with respect to the second, into the dual construction for the third.)

6·4 Conjugate points and conjugate lines.* The following 'harmonic property' can be traced back to Apollonius:

6·41 *Any two conjugate points on a secant PQ are harmonic conjugates with respect to P and Q.*
Proof: The self-conjugate points P and Q are the invariant points of the involution of conjugate points on the line PQ (see 4·63).
    Dually,

6·42 *Any two conjugate lines through an exterior point are harmonic conjugates with respect to the tangents that can be drawn through the point.*

The following theorem will enable us to construct the polar of a given point with respect to a given conic:

6·43 *If a quadrangle is inscribed in a conic, its diagonal triangle is self-polar.*
Proof: Let the diagonal points of the inscribed quadrangle $PQRS$ be

$$A = PS \cdot QR, \quad B = QS \cdot RP, \quad C = RS \cdot PQ,$$

as in Fig. 6·4A. The line BC meets the sides QR and PS in points A_1 and A_2 such that $H(QR, AA_1)$ and $H(PS, AA_2)$. By 6·41, A_1 and A_2 are

---

* von Staudt 1847, pp. 139–40, §§ 249–51.

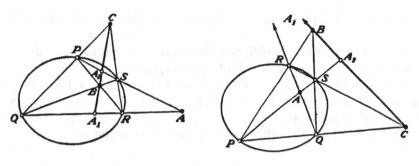

Fig. 6·4A

conjugate to A. Hence the line BC, which joins them, is the polar of A. Similarly, CA is the polar of B and AB of C.

6·44 *To construct the polar of a given point A, not on the conic, draw any two secants QR and PS through A; then the polar is*

$$(QS \cdot RP)(RS \cdot PQ).$$

In other words, we draw two secants through A to form an inscribed quadrangle with diagonal triangle ABC, and then the polar of A is BC.

The dual construction presupposes that we know the tangents from any given exterior point. This presents no serious difficulty (since their points of contact lie on the polar of the given point); but the tangents

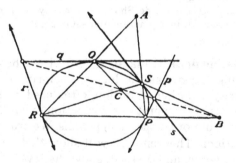

Fig. 6·4B

are not immediately apparent, for the simple reason that we are in the habit of dealing with loci rather than envelopes. If we insist on regarding the conic as a locus, we may construct the pole of a given line as the point of intersection of the polars of any two points on the line. Then:

6·45 *To construct the tangent at a given point A on the conic, join A to the pole of an arbitrary line through A.*

Theorem 6·43 and its dual may be neatly brought together as follows:

6·46 *A quadrangle PQRS, inscribed in a conic, has the same diagonal triangle as the quadrilateral of tangents pqrs.*

Proof: Defining A, B, C as before, we observe that A lies on the polars of both $q \cdot r$ and $p \cdot s$ (see Fig. 6·4B). Hence the diagonal $(q \cdot r)(p \cdot s)$ of the quadrilateral is the polar of A, that is, it coincides with BC. Similarly the other two diagonals are CA and AB.

EXERCISES

1. Let B and C be two conjugate points with respect to a given conic. Let an arbitrary line through C meet the conic in P and Q, while BP and BQ meet the conic again in R and S, respectively. Prove that C, R, S are collinear.

2. If PQR is a triangle inscribed in a conic, show that infinitely many self-polar triangles can be found having one vertex on each side of PQR.

3. Show that infinitely many triangles can be inscribed in a given conic in such a way that each side passes through one vertex of a given self-polar triangle.

4. Prove that a conic is transformed into itself by any harmonic homology whose centre is the pole of its axis.

5. If the six points P, P', Q, Q', R, R' of the Desargues configuration (Fig. 2·2A) lie on a conic, prove that O is the pole of the line ABC with respect to that conic. Deduce that the homology of 5·24 is then harmonic, so that the polarities $(PQR)(Oo)$ and $(P'Q'R')(Oo)$ coincide. Show also that any line through O meets the six sides of the two triangles in a quadrangular set of points.*

6·5 Two possible definitions for a conic.† We have followed von Staudt in defining a conic as the locus of self-conjugate points in a hyperbolic polarity. Another definition, often used, is Steiner's (1832): A conic is the locus of the point of intersection of corresponding lines of two projective (but not perspective) pencils. We proceed to reconcile these two definitions. Theorem 6·52 will show that every von Staudt conic is a Steiner conic, and 6·54 will show that every Steiner conic is a von Stuadt conic.

As a first step we need:

6·51 *Seydewitz's theorem. If a triangle is inscribed in a conic, any line conjugate to one side meets the other two sides in conjugate points.*

Proof: Consider an inscribed triangle PQR. Any line c conjugate to PQ is the polar of some point C on PQ. Let RC meet the conic again in

---

* Mathews 1914, p. 338, Ex. 104.

† von Staudt 1847, pp. 141–4, §§ 253–8.

S, as in Fig. 6·4A. By 6·43, c joins the points

$$A = PS \cdot QR, \quad B = QS \cdot RP.$$

These conjugate points A and B are the intersections of c with the sides QR and RP of the given triangle.

6·52 Steiner's theorem. *Let lines x and y join a variable point on a conic to two fixed points on the same conic; then $x \barwedge y$.*

Proof: The tangents p and q, at the fixed points P and Q, intersect in D, the pole of PQ. Let c be a fixed line through D (but not through P or Q), meeting x and y in B and A, as in Fig. 6·5A. By 6·51, BA is a pair of

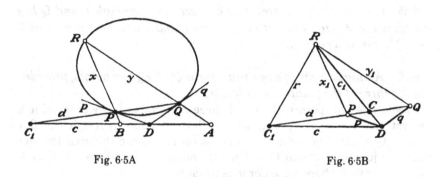

Fig. 6·5A   Fig. 6·5B

the involution of conjugate points on c. Hence, when the point $x \cdot y$ varies on the conic,

$$x \barwedge B \barwedge A \barwedge y.$$

In particular, when x is p, B is D, A is on $d = PQ$, and y is d; and when x is d, B is on d, A is D, and y is q,

We shall prove the converse theorem with the help of the following lemma:

6·53 *A conic is determined when three points on it and the tangents at two of these are given.*

Let P, Q, R be the given points, PD and QD the given tangents, and C_1 the harmonic conjugate of $C = PQ \cdot RD$ with respect to P and Q, as in Fig. 6·5B. Consider the definite correlation that transforms the four points P, Q, R, D into the four lines PD, QD, RC_1, PQ and consequently transforms PQ into D, RD into C_1, and C into $C_1 D$. This induces in PQ a projectivity $PQC \barwedge PQC_1$. Since $H(PQ, CC_1)$, this is the involution $PQCC_1 \barwedge PQC_1 C$; hence the correlation transforms C_1 into CD and is a polarity (by 5·62 applied to triangle DCC_1). Since the polars of P, Q, R are PD, QD, RC_1, the polarity determines a conic having the desired properties.

6·54   Steiner's construction. *Let variable lines x and y pass through fixed points P and Q in such a way that x ⊼ y but not x ⊼̄ y; then the locus of x · y is a conic through P and Q.*

Proof: Since the projectivity x ⊼ y is not a perspectivity, the line d = PQ does not correspond to itself. Hence there exist lines p and q such that the projectivity relates p to d and d to q. By 6·53 we can find a conic touching p at P, q at Q and passing through any other particular position $x_1 · y_1$ of the variable point x · y. By 6·52 the conic determines a projectivity between pencils through P and Q; and this must coincide with the given projectivity x ⊼ y, since it relates the three lines x_1, p, d through P to the three lines y_1, d, q through Q.

6·55   Corollary. *If a projectivity between lines through P and Q has the effect xpd ⊼ ydq, where d is PQ, then p and q are the tangents at P and Q to the locus of x · y.*

6·56   *A unique conic can be drawn through five given points, provided that no three of the points are collinear.*

Proof: The two points P, Q and three positions of x · y determine a projectivity $x_1 x_2 x_3$ ⊼ $y_1 y_2 y_3$, which yields a conic through the five points (by 6·54). Conversely, if a point on any conic through the five points is joined to P and Q by lines x and y, we must have $x x_1 x_2 x_3$ ⊼ $y y_1 y_2 y_3$ (by 6·52); hence the conic is unique.

(Lemma 6·53 may be regarded as the special case that arises when two pairs of the five given points coincide.)

The duals of 6·51 and 6·54 are sufficiently important to be stated explicitly:

6·57   *If a triangle is circumscribed about a conic, any point conjugate to one vertex is joined to the other two vertices by conjugate lines.*

6·58   *Let points X and Y vary on fixed lines p and q in such a way that X ⊼ Y but not X ⊼̄ Y; then the envelope of the line XY is a conic touching p and q.* *

It follows from 6·53 that infinitely many conics can be drawn to touch two fixed lines at two fixed points. Such conics are said to have *double contact* (with one another).

6·59   *Of the conics that touch two given lines at given points, those which meet a third line (not through either of the points) do so in pairs of an involution.*

Proof: Let such a conic touch AB at P, AC at Q, and meet BC in R and S, as in Fig. 6·5C. Let the polar of the point

---

* Proved by Chasles in 1828.

$$M = PQ \cdot BC$$

meet PQ in L and BC in N. Then by 6·41 we have H(RS, MN), H(PQ, ML), and consequently H(BC, MN). Hence RS is a pair of the hyperbolic involution $(MM)(BC)$, whose invariant points are M and N.

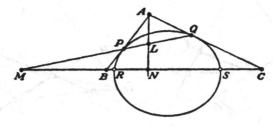

Fig. 6·5C

EXERCISES

1. Dualize 6·52 and 6·53.

2. If variable points X and Y on fixed lines p and q are conjugate for a given polarity, while the point $p \cdot q$ is not self-conjugate, prove that the line XY envelops a conic.*

3. Show that the four points P, Q, R, S of 6·46 lie on a conic through $p \cdot s$ and $q \cdot r$.†

4. Let A and B be a pair of conjugate points with respect to a conic, and let PQ be a secant conjugate to AB. Prove that AQ and BP meet on the conic. (*Hint: BP* meets the conic again in the point R of §6·4, Ex. 1.)

5. Consider a variable conic inscribed in a given quadrilateral. Show that the line joining its points of contact with two sides of the quadrilateral passes through a fixed point. Identify this point.‡

6. Measure off points X_0, X_1, ..., X_5 at equal intervals along a line and Y_0, Y_1, ..., Y_5 similarly along another line through $X_5 = Y_0$. The joins $X_n Y_n$ visibly envelop a conic, in accordance with 6·58. (Of course, this method of setting up a projectivity is not playing the game. We shall have to wait until Chapter 8 to see why it is valid and why the conic is a parabola.)

6·6 Construction for the conic through five given points.
The following construction was given by Braikenridge in 1733, but

* Chasles 1865, pp. 10, 137.
† Chasles 1865, p. 138.
‡ Graustein 1930, p. 325.

his priority was contested by Maclaurin in a rather disagreeable controversy:

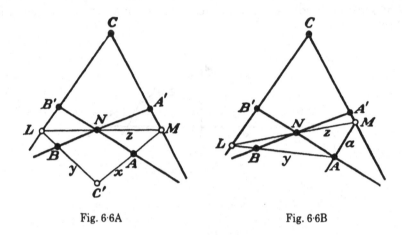

Fig. 6·6A   Fig. 6·6B

6·61   *If the sides of a variable triangle pass through three fixed non-collinear points, while two vertices run along fixed lines, the third vertex will trace a conic through two of the given points.*

Proof: Let LMC' be the variable triangle whose sides $x = MC'$, $y = LC'$, $z = LM$ pass through fixed points A, B, N, while the vertices L and M run along fixed lines CB' and CA', as in Fig. 6·6A. Then

$$x \overset{CA'}{\barwedge} z \overset{CB'}{\barwedge} y.$$

The projectivity $x \barwedge y$ could be a perspectivity only if N lay on AB. By 6·54, the locus of $C' = x \cdot y$ is a conic through A and B.

This conic passes also through A' (on NB), B' (on NA), and C. For when z coincides with NB, y does the same, while M and C' coincide with A'. Similarly, when z coincides with NA, C' coincides with B'. Finally, when z coincides with NC, the points L, M, C' all coincide with C.

In other words, given five points A, B, C, A' B' of which no three are collinear, we may construct any number of positions of a sixth point C' on the conic $ABCA'B'$ in the following manner: Through the point $N = AB \cdot BA'$ draw an arbitrary line z, meeting CB' in L and CA' in M. Then $C' = AM \cdot BL$ is another point on the conic.

By 6·55 the tangent at A (Fig. 6·6A) is the position taken by x when y coincides with BA, so that L is $AB \cdot CB'$, as in Fig. 6·6B. Hence:

6·62   *The tangent at A to the conic $ABCA'B'$ is the line AM where* $M = CA' \cdot (AB' \cdot BA')(AB \cdot CB')$.

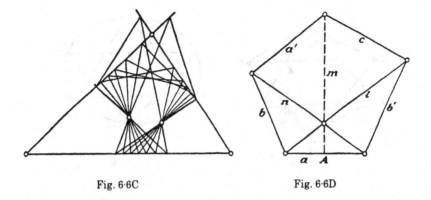

Fig. 6·6C   Fig. 6·6D

EXERCISES

1. Dualize 6·61, and deduce a construction for any number of tangents to the conic determined by five given lines, no three concurrent (see Fig. 6·6C).

2. Dualize 6·62 to obtain a construction for the point of contact of any one of five given tangents (see Fig. 6·6D).

3. Observe that Fig. 4·8A is the same as Fig. 6·6A if we name the points of the former as follows:

$$A = a \cdot b, \quad B = a' \cdot b', \quad C = c \cdot c', \quad A' = a \cdot a', \quad B' = b \cdot b', \quad C' = x \cdot x',$$

$$L = s \cdot x'. \quad M = r \cdot x, \quad N = a' \cdot b.$$

4. Prove that, if the three pairs of opposite sides of a hexagon meet in three collinear points, then the six vertices lie on a conic or on two lines. (*Hint:* Consider the hexagon $AB'CA'BC'$ of Fig. 6·6A.) Dually, if the three diagonals of a hexagon are concurrent, the six sides touch a conic or pass through two points.

5. Given a pair of Desargues triangles, show that the six points in which the sides of one triangle meet the noncorresponding sides of the other lie on a conic or on two lines. Dually, the six lines joining the vertices of one triangle to the non-corresponding vertices of the other are tangents of a conic or pass through two points.

6·7 Two triangles inscribed in a conic.* The following results are interesting in themselves, besides illustrating the use of Steiner's construction.

6·71 *If two triangles are self-polar for a given polarity, their six vertices lie on a conic or on two lines.*

___
* von Staudt 1847, pp. 174–5, §§ 299, 300.

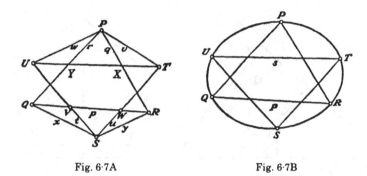

Fig. 6·7A   Fig. 6·7B

Proof: The given self-polar triangles PQR and STU determine further points and lines

$$V = p \cdot t, \quad W = p \cdot u, \quad v = PT, \quad w = PU, \quad x = QS, \quad y = RS,$$

as in Fig. 6·7A. With the help of 2·71, we find

$$rqvw \barwedge RQVW \barwedge \barwedge QRWV \barwedge xyut.$$

Therefore, by 6·54, the four points Q, R, T, U lie on a conic through P and S, unless a vertex of one triangle lies on a side of the other; but if S lies on QR, its polar TU passes through the pole P, and the six vertices lie by threes on two lines.

Dually, the six sides either touch a conic or pass by threes through two points (say P and S).

In §5·9 we considered two kinds of degenerate polarity, which essentially amounted to an involution of lines through a point and an involution of points on a line. When these involutions are hyperbolic, their 'self-conjugate' points and lines yield *degenerate conics*: in the former case a pair of lines, and in the latter a pair of points. This idea enables us to delete the clause '*but not $x \barwedge y$*' from 6·54, which may now be expressed as follows: *A conic is the locus of a point that is joined to two fixed points by corresponding lines of two projectively related pencils.* For although when $x \barwedge y$ the locus of $x \cdot y$ is strictly only one line, we must admit that any point collinear with A and B shares with the points of that locus the property of being joined to A and B by corresponding lines of the related pencils.

Finally, the two kinds of doubly degenerate polarity (§5·9) yield two *doubly degenerate conics*: a line (usually referred to as 'two coincident lines') and a point (or 'two coincident points').

Given a polarity $(PQR)(Ss)$, we observe that the involution of conjugate points on s is the quadrangular involution determined on s by the quadrangle $PQRS$; for instance, PS meets s in a point conjugate to $p \cdot s$. Let TU be any pair of this involution. Then STU is a second self-polar triangle. By 6·71, the six points lie on a conic, possibly degenerate. Conversely (Fig. 6·7B), U may be described as the second intersection of s with the conic $PQRST$, where T is an arbitrary point on s. Hence:

6·72 Desargues's involution theorem. *Of the conics that can be drawn through the vertices of a given quadrangle, those which meet a given line (not through a vertex) do so in pairs of the quadrangular involution.*

6·73 *If two triangles have six distinct vertices, all lying on a conic, they are self-polar for some polarity.*

Proof: Triangles PQR and STU, inscribed in a conic, are self-polar for the polarity $(PQR)(Ss)$, where $s = TU$.

EXERCISES

1. Observe how 6·71 and 6·73 serve to establish the following theorem, due to Steiner: *If two triangles are inscribed in a conic, their six sides touch a conic (and conversely).*

2. Prove that a given line touches at most two of the conics through P, Q, R, S.

3. The formal self-duality of the Desargues configuration (Figs. 2·2A, B) continues to hold if we interchange the names of P and P', Q and Q', R and R'. Prove that this *duality* cannot be realized as a *polarity* (cf. 5·71) unless these six points lie on a conic (as in §6·4, Ex. 5).

6·8 Pencils of conics. We saw in §5·8 that the polarities $(ABC)(Pp)$, where p turns about a fixed point P', have the property that the polars of an arbitrary point X all pass through a corresponding point X'. In other words, the pencil of polarities determines a correspondence $X \to X'$ between points that are simultaneously conjugate for all the polarities. Now, if X runs along a fixed line o, how does X' behave? The answer will emerge while we are proving the following theorem:

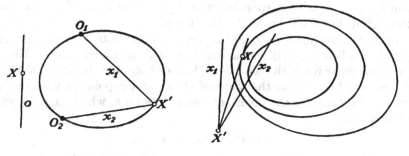

Fig. 6·8A   Fig. 6·8B

6·81   *The locus of poles of a fixed line with respect to a pencil of polarities is a conic or a line.*

Proof: By the dual of 5·81 the locus is a line if the pencil is a 'self-dual system'; but we are more interested in the general case. Let O_1 and O_2 be the poles of the given line o with respect to two particular polarities in the pencil (see Fig. 6·8A). When the point X varies on o, its two polars $x_1 = O_1 X'$ and $x_2 = O_2 X'$ rotate about the fixed points O_1 and O_2 and we have

$$x_1 \barwedge X \barwedge x_2.$$

Three particular positions for X are where o cuts the sides of the common self-polar triangle ABC of the polarities; then the conjugate points X' are the vertices of that triangle, in turn. Hence the locus of X' (while X runs along o) is the conic $ABCO_1 O_2$, which is nondegenerate provided that no three of these five points are collinear. Each position of X' depends only on X and the pencil, not on the two selected polarities. Hence any other polarity in the pencil will give a pole O on this same conic. In other words, the conic is both the locus of X' (when X varies on o) and the locus of the pole O (when the polarity varies in the pencil). The vertices of the triangle occur as positions of O when we take p to be $P'A$, $P'B$, or $P'C$, as in §5·9.

An alternative procedure is to construct O as the intersection of the polars of two fixed points W and X on o. By 5·81, these polars, w, and x, describe pencils through two fixed points W' and X'. In the notation of the proof of that theorem,

$$x \barwedge Y \barwedge E \barwedge p,$$

and similarly $w \barwedge p$, whence $w \barwedge x$. Thus the locus of O is the conic $ABCW'X'$.

We still have to investigate the possibility that this conic degenerates. We shall find that, when it degenerates into two lines, the loci of X' and O no longer coincide but X' runs along one line and O along the other.

On possibility is that O_1 might lie on a side of triangle ABC, say a. This can happen only if the given line o passes through A, in which case not only O_1 but also O_2 lies on a. The position A for X makes x_1 and x_2 coincide with a; therefore $x_1 \overline{\barwedge} x_2$, and the locus of X' is a line through A; but the locus of O is the line a.

On the other hand, if O_1 and O_2 are collinear with a vertex, say A (but do not lie on a side), we have $AO_1 = AO_2$, so that the point $a \cdot o$ has the same polar for both polarities. Thus $x_1 \overline{\barwedge} x_2$, and the locus of X' is a line, viz. BC. To find the locus of O, we let the polarity vary in its pencil and consider the polars of two points on o, which may conveniently be taken to be

$$W = b \cdot o \quad \text{and} \quad X = c \cdot o.$$

The relation $w \barwedge x$ is such that, when p is $P'A$ (so that the polarity is degenerate), w is BA and x is CA. Since the locus of $O = w \cdot x$ includes three collinear points A, O_1, O_2, we have $w \; \overline{\wedge} \; x$, and the locus consists of the line AO_1. This is the case of a self-dual system of polarities (see 5·82).

Here we have assumed that o does not pass through the special vertex A; but if it does, its pole O is the same for all the polarities and X' constantly coincides with O.

6·82 *Every pencil of polarities determines a pencil of conics: one conic through each point of general position.*

Proof: Consider once more the polarities $(ABC)(Pp)$, with p through the fixed point P'. Since one possible position for p is $P'p$, one of the polarities determines a conic that touches PP' at P. In fact, every point X lies on such a conic (touching XX' at X), though for some positions of X the conic may be degenerate.

In particular, a self-dual system of polarities (see 5·82) determines a *self-dual system of conics* (e.g. Fig. 6·8B or C), including the line a and point A as doubly degenerate members.

It may happen that two conics of a pencil have a common point. Then this point is self-conjugate for two, and therefore all, of the polarities.

If such a common point P lies on the side a of the common self-polar triangles ABC, then all the conics have the same tangent PA at P and the same tangent QA at Q, the harmonic conjugate of P with respect to B and C (see Fig. 6·8C). Thus, *if two conics of a self-dual system have a*

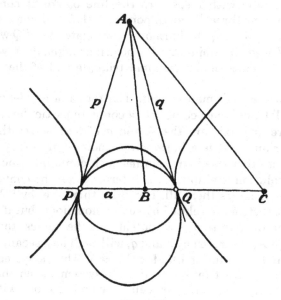

Fig. 6·8C

common point, all of them have double contact. This system of conics includes the line-pair p, q as a degenerate member of the first kind and the point-pair P, Q as a degenerate member of the second kind (as well as the point $p \cdot q$ and the line PQ as doubly degenerate members).

On the other hand, if the common point P does not lie on a side of triangle ABC, then we can find (as in 2·42) a definite quadrangle $PQRS$ having ABC for its diagonal triangle. All four vertices of this quadrangle are self-conjugate; for example, Q (being the harmonic conjugate of P with respect to the conjugate points C and C_1 of Fig. 2·4C) is the second invariant point of the involution of conjugate points on CP. Hence in this case the pencil consists of *all the conics circumscribed to a quadrangle* (as in 6·72), and we call it a *quadrangular* pencil of conics. The quadrangular pencil includes three degenerate conics: the pairs of opposite sides of the quadrangle.

The 'double-contact' system may be regarded as a limiting case of this quadrangular pencil, when the points P, Q, R, S approach coincidence in pairs, PS and QR, in such a way that the lines PS and QR have definite limiting positions p and q. For then the conics will all touch p at P and q at Q.

As a special case of 6·81, we have the following:

6·83   *Given a quadrangular pencil of conics and a line not through a diagonal point of the quadrangle, the locus of poles of the line is a conic.*

This conic, which is not only the locus of poles O of the given line o but also the locus of the point X', which is conjugate to a variable point X on o with respect to all the conics, is called the *nine-point conic* of the quadrangle with respect to the line o. For it contains nine special points: the three diagonal points A, B, C and one further point on each side; e.g. on PQ the harmonic conjugate of $o \cdot PQ$ with respect to P and Q. These nine points are the positions taken by X' when X lies, in turn, on the sides of the diagonal triangle and of the quadrangle itself.

Since conics are ultimately defined in terms of incidence, any collineation will transform a conic into a conic. In particular, a homology whose centre and axis are the A and a of 5·82 leaves the self-dual system invariant; if it is a harmonic homology, it leaves each conic invariant, but otherwise it transforms each conic into another. If the common involution is elliptic, A is interior to all the conics and each line through A meets them all. Hence in this case, by 5·21, any two conics of the system are related by such a homology; but if the involution is hyperbolic, so that A is outside all the conics, then any two lines through A, separated by p and q, will each be a secant for some of the conics and an exterior line for the rest. Thus a system of conics having double contact falls into two subsystems, such that any two conics of the same subsystem are related by a homology with centre A

and axis a. (These subsystems are represented by ellipses and hyperbolas in Fig. 6·8C, but of course that distinction belongs to affine geometry; in projective geometry they are exactly alike.) On the other hand, each line through P meets all the conics (of both subsystems); hence any two conics of the whole system are related by a homology with centre P and axis q (and likewise by one with centre Q and axis p). In either case:

6·84 *Any two conics of a self-dual system are related by a homology.*

EXERCISES

1. Theorem 6·82 shows that every pencil of polarities includes infinitely many hyperbolic polarities. Prove that the pencil consists entirely of hyperbolic polarities if a common pair of conjugate points (P and P') both lie in the same one of the four regions determined by the common self-polar triangle ABC; but infinitely many elliptic polarities occur as well if such conjugate points lie in different regions. (In the case of a self-dual system the same distinction depends on whether the common involution is hyperbolic or elliptic.)

2. In the presence of a general pencil of polarities, any line o determines a conic (the locus of poles of o; se 6·81). Show that a pencil of lines o determines a pencil of such conics.

3. If two quadrangles have the same diagonal points, prove that either they share a pair of opposite sides or their eight vertices lie on a conic. (*Hint:* Consider the conic determined by one quadrangle and one vertex of the other.)

4. Using 6·46 and the dual of Ex. 3, prove that, if two conics intersect in four points, the eight tangents at these points either pass by fours through two points or touch a conic (*Salmon's conic*).

5. For the nine-point conic show that three pairs of the nine points (e.g. the harmonic conjugate of $o \cdot PQ$ on PQ and that of $o \cdot RS$ on RS) are joined by lines conjugate to o.

CHAPTER 7

Projectivities on a Conic

This chapter deals with those properties of a non-degenerate conic which may be most readily derived by means of the notion that the points on the conic form a range, resembling in many ways the points on a line. Pascal's theorem is the most famous instance; but its original proof must have been different. The idea of projectivity on a conic is due to Bellavitis (1838). We shall see that the construction for such a projectivity is simpler than for a projectivity on a line. In fact, some authors, such as Holgate, rearrange the material so as to treat ranges on a conic before ranges on a line. Involutions are especially easy to deal with, for the joins of pairs of corresponding points are concurrent, as we shall see in § 7.5.

7·1 Generalized perspectivity. Steiner's theorem (our 6·52) enables us to regard the points on a conic as a *range* that can be related to an ordinary range or pencil or to another range on the same (or another) conic. Thus, if variable points R and R' on a conic are joined to a fixed point *on the same conic* by lines x and x', where $x \barwedge x'$, then we are justified in writing

$$R \barwedge R',$$

since the fixed point P could just as well be replaced by another such point Q (on the same conic), as in Fig. 7·1A.

If x meets a fixed line (not through P) in B, we write

$$R \overset{P}{\barwedge} B$$

and call this relation a *generalized perspectivity*. (Always remember that P has to lie on the conic.) When R coincides with P, x is the tangent there (by 6·55) and B is where this tangent meets the fixed line.

This notion makes it easy to prove the following theorem:

7·11 *If the six vertices of two triangles lie on a conic, the six sides touch a conic (and conversely).*

Proof: The given inscribed triangles *PQR* and *STU* determine further points

$$V = QR \cdot SU, \quad W = QR \cdot ST, \quad X = PR \cdot TU, \quad Y = PQ \cdot TU$$

(see Fig. 6·7A), and we have

$$RQVW \overset{S}{\barwedge} RQUT \overset{P}{\barwedge} XYUT.$$

Since *P* and *S* are distinct, the projectivity $RQVW \barwedge XYUT$ is not a perspectivity. Therefore, by 6·58, the lines *RX*, *QY*, *VU*, *WT* are tangents of a conic that also touches *QR* and *UT*. (The converse is the dual theorem. For another proof, see §6·7, Ex. 1.)

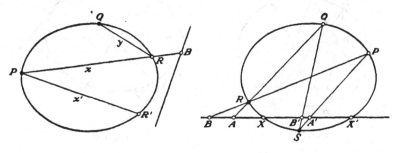

Fig. 7·1A　　　　　　　　　　Fig. 7·1B

As another application of this method, here is an alternative proof*
of Desargues's involution theorem (our 6·72). Let the given line meet
the sides *QR*, *PS*, *RP*, *QS* of the quadrangle, and one of the conics, in
the points *A*, *A'*, *B*, *B'*, *X*, and *X'*, as in Fig. 7·1B. Then

$$ABXX' \overset{R}{\barwedge} QPXX' \overset{S}{\barwedge} B'A'XX'.$$

Hence, by 4·66 and 4·71, *XX'* is a pair of quadrangular involution
$(AA')(BB')$.

This proof has the advantage of remaining valid when *P* coincides
with *S* (or *Q* with *R*) so that the conics form a 'contact' pencil instead
of a quadrangular pencil. It even remains valid when *P = S* and *Q = R*,
so that the conics form a 'double-contact' system, as in 6·59. In each
case, those conics of the pencil which meet a line of general position do
so in pairs of an involution.

EXERCISE

Establish the relation $PQRS \barwedge QPSR$ for any four points on a conic. (*Hint:* Use
2·71.)

---

* von Staudt 1847, p. 176, §301.

7·2 Pascal and Brianchon. It is interesting to observe the simplicity of the projective approach to Pascal's theorem in comparison with any proof that could possibly have occurred to Pascal himself (see § 1·7). A variant of the following proof could be obtained by regarding this theorem as the converse of Braikenridge's construction (our 6·61).

7·21 Pascal's theorem. *If a hexagon is inscribed in a conic, the three pairs of opposite sides meet in three collinear points.*

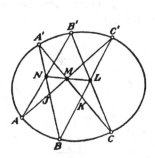
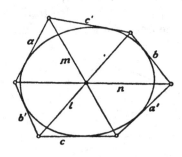

Fig. 7·2A   Fig. 7·2B

*Proof:** Let $AB'CA'BC'$ be the hexagon, so that the points to be proved collinear are

$$L = BC' \cdot CB', \quad M = CA' \cdot AC', \quad N = AB' \cdot BA',$$

as in Fig. 7·2A. Using further points

$$J = AC' \cdot BA' \quad \text{and} \quad K = BC' \cdot CA',$$

we have

$$A'NJB \overset{A}{\barwedge} A'B'C'B \overset{C}{\barwedge} KLC'B.$$

Thus B is an invariant point of the projectivity $A'NJ \barwedge KLC'$, which accordingly is a perspectivity, namely, $A'NJ \overset{M}{\barwedge} KLC'$ (since the joins $A'K$ and JC' pass through M). Hence NL passes through M.

Note the close analogy with Pappus's theorem (our 4·31), which is Pascal's theorem as applied to a degenerate conic. (Some German authors call Pappus's theorem 'Pascal's theorem'.) Anyone who dislikes the generalized perspectivity may deduce

$$A'NJB \barwedge KLC'B$$

from the observation that the lines AA', AB', AC', AB are projectively related to the respective lines CA', CB', CC', CB.

---

* von Staudt 1847, p. 143, § 257; Enriques 1904, p. 225.

The dual of Pascal's theorem was discovered by Brianchon, nearly 170 years later.

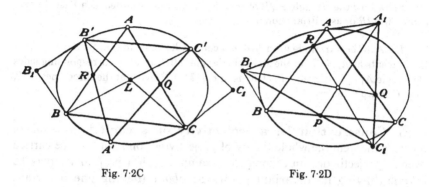

Fig. 7·2C   Fig. 7·2D

7·22 Brianchon's theorem. *If a hexagon is circumscribed about a conic, the three diagonals are concurrent* (see Fig. 7·2B).

The following is typical of many applications of Pascal's theorem:

7·23 *If A, B, C, A', B', C' are six points on a conic, while the tangents at these points are a, b, c, a', b', c', then the three lines $(a \cdot a')(BC \cdot B'C')$, $(b \cdot b')(CA \cdot C'A')$, $(c \cdot c')(AB \cdot A'B')$ are concurrent.*

Proof: Let the various points be named

$$A_1 = a \cdot a', \qquad B_1 = b \cdot b', \qquad C' = c \cdot c',$$

$$L = BC' \cdot CB', \quad M = CA' \cdot AC', \quad N = AB' \cdot BA',$$

$$P = BC \cdot B'C', \quad Q = CA \cdot C'A', \quad R = AB \cdot A'B',$$

as in Fig. 7·2C. From the Pascal hexagon $CABC'A'B'$, the three points L, Q, R are collinear. By 6·43, L lies on the polar of $BB' \cdot CC'$, which is $B_1 C_1$. Thus L lies on both $B_1 C_1$ and QR. Similarly, M lies on $C_1 A_1$ and RP, N on $A_1 B_1$ and PQ. But L, M, N are collinear (from the Pascal hexagon $AB'CA'BC'$). Hence $A_1 B_1 C_1$ and PQR are Desargues triangles.

EXERCISES

1. Verify that $A'RSBXX'$ in Fig. 7·2D is a Brianchon hexagon.

2. Show how the pentagon $AB'CA'B$ of 6·62 may be regarded as a limiting case of the Pascal hexagon when C' coincides with A.

3. Show how the quadrangle $PQSR$ of 6·46 may be regarded as a limiting case of the Pascal hexagon when two vertices coincide at Q and two others at R, so that the 'hexagon' is $PQQSRR$. Similarly, the quadrilateral $prqs$ may be regarded as a limiting Brianchon hexagon $pprqqs$.

4. Let a conic be defined by four points and the tangent at one of them. Construct the tangent at another one of the four points.

5. Show how the triangle PQR of §6·3, Ex. 3, may be regarded as a Pascal hexagon $PPQQRR$ or as a Brianchon hexagon $ppqqrr$.

6. Let ABC be a triangle inscribed in a conic. Choosing any points A_1, B_1, C_1 on the tangents at A, B, C, let the sides of triangle ABC meet the corresponding sides of triangle $A_1B_1C_1$ in points P, Q, R, as in Fig. 7·2D. Prove that the three lines A_1P, B_1Q, C_1R are concurrent.*

7·3 Construction for a projectivity on a conic.† In virtue of Steiner's theorem the whole theory of projectivities on a line can be carried over to projectivities on a conic; for example, such a projectivity may be *elliptic* (having no invariant point), *parabolic* (having one invariant point), or *hyperbolic* (having two invariant points), and in the last case it is either *direct* or *opposite* (according as it preserves or reverses the sense around the conic). The construction for the transform of a given point is actually easier than in 2·72, for now we can make use of an auxiliary line called the axis of the projectivity.

The *axis* of the projectivity $ABC \barwedge A'B'C'$ on a conic is the Pascal line of the hexagon $AB'CA'BC'$, namely, the line o determined by any two of the three points $BC' \cdot CB'$, $CA' \cdot AC'$, $AB' \cdot BA'$.

7·31 *Given seven points A, B, C, X, A', B', C' on a conic, we can locate the point X' such that $ABCX \barwedge A'B'C'X'$ as the second intersection of the conic with the line AF, where F is $XA' \cdot o$.*
Proof: If AA' meets the axis o in G (Fig. 7·3A), we have

$$ABCX \overset{A'}{\barwedge} GNMF \overset{A}{\barwedge} A'B'C'X'.$$

Thus the *cross joins* of any two pairs of corresponding points meet on the axis.

The general projectivity between ranges on two distinct lines (Fig. 4·3C) may be regarded as a degenerate case, the Pascal line of the conic becoming the Pappus line of the line pair.

7·32 *A projectivity on a conic is determined when its axis and one pair of corresponding points are given.*
Proof: Given the axis and AA', we have $X \overset{A'}{\barwedge} F \overset{A}{\barwedge} X'$. In other words, for any X on the conic, there is a unique X' such that $AX' \cdot XA'$ lies on the axis.

---
* This theorem, due to Kenneth Leisenring, was put into the more manageable form 7·23 by Alex Rosenberg.
† Enriques 1904, p. 251.

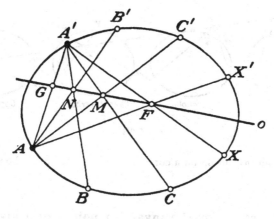

Fig. 7·3A

The only way in which X' can coincide with X is by X lying on the axis. Hence:

7·33 *The invariant points (if any) of a projectivity on a conic are the common points of the axis and the conic.*

7·34 *The projectivity is elliptic, parabolic, or hyperbolic, according as its axis is an exterior line, a tangent, or a secant.*

A secant decomposes the conic into two arcs. Hence, by 4·42, a hyperbolic projectivity is direct or opposite according as two corresponding points lie both on the same arc or one on each. The four types of projectivity are illustrated in Fig. 7·3B.

7·35 *Any projectivity on a conic determines a collineation of the whole plane.*

Proof: By 6·53, a unique conic can be drawn to touch two given lines a and b at two given points A and B and to pass through another given point C. Consider the quadrangle $ABCD$, where D is $a \cdot b$, and another such quadrangle $A'B'C'D'$ (determining the same conic). These two quadrangles are related by a unique collineation (see 5·12), which preserves the conic and induces the projectivity $ABC \barwedge A'B'C'$ on it.

EXERCISES

1. Show that an involution on a conic determines a harmonic homology of the whole plane.

2. Show that any collineation leaving a conic invariant is expressible as the product of two harmonic homologies. (*Hint:* Use 4·69.)

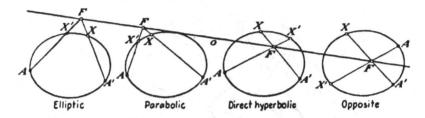

Fig. 7·3B

3. Show that an involution on a conic determines a harmonic homology of the whole plane.

7·4 Construction for the invariant points of a given hyperbolic projectivity. Given a hyperbolic projectivity on a conic, we can easily locate its invariant points by drawning its axis (see 7·33). This suggests the following construction (due to Steiner) for the invariant points of a given hyperbolic projectivity $ABC \,\overline{\wedge}\, A'B'C'$ on a line:

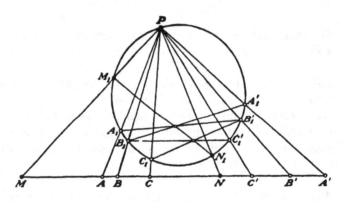

Fig. 7·4A

Draw any conic (in practice, a circle*), and project the given points from any point P on the conic into $A_1, B_1, C_1, A'_1, B'_1, C'_1$, on the conic, as in Fig. 7·4A. Draw the axis

$$(A_1 B'_1 \cdot B_1 A'_1)(B_1 C'_1 \cdot C_1 B'_1)$$

to meet the conic in M_1 and N_1. Project these points from P back onto the original line, and we obtain the desired invariant points M and N.

Proof: $MNABC \,\overset{P}{\overline{\wedge}}\, M_1 N_1 A_1 B_1 C_1 \,\overline{\wedge}\, M_1 N_1 A'_1 B'_1 C'_1 \,\overset{P}{\overline{\wedge}}\, MNA'B'C'$.

---

* The familiar process of reciprocation with respect to a circle is an instance of a polarity; therefore a circle is a conic. We shall return to this subject in §9·2.

What will happen if we try to carry out this construction when the given projectivity is not hyperbolic?

7·5 Involution on a conic.* The involution $(AA')(BB')$ may be regarded as the special case of the projectivity

$$ABC \barwedge A'B'C'$$

that arises when $C = B'$ and $C' = B$, as in Fig. 7·5A. The axis

$$o = (AB' \cdot BA')(AC' \cdot CA') = (AB' \cdot BA')(AB \cdot A'B')$$

is one side of the diagonal triangle of the quadrangle $AA'BB'$, and its pole is the opposite vertex $O = AA' \cdot BB'$ of that triangle. Hence:

7·51 *The pairs of an involution of points on a conic are joined by concurrent lines*; i.e. *they are cut out by a pencil of secants.*

Conversely, any pencil of secants determines an involution, namely, $(AA')(BB')$, where AA' and BB' are any two of the secants.

The axis, being the polar of O, contains the point of intersection of the tangents at any two corresponding points. This could also be inferred by considerations of continuity; for the tangents at A and A' are the limiting positions of the cross joins of the pairs AA' and $B'B$ when B approaches A.

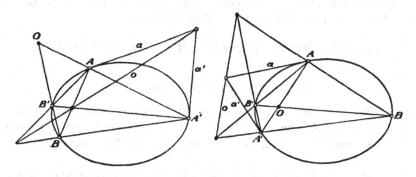

Fig. 7·5A

The point $O = AA' \cdot BB'$ is called the *centre* of the involution $(AA')(BB')$. Since it is the pole of the axis, 7·34 implies the following:

---

* Cremona 1960, p. 160. For an alternative treatment see Enriques 1930, pp. 256–9) or Veblen and Young (1910, p. 222).

7·52   *The centre O of an involution is interior or exterior according as the involution is elliptic or hyperbolic, and in the latter case the invariant points are the points of contact of the two tangents that can be drawn through O.*

7·53   Corollary. *Four points on a conic satisfy AA'||BB' if and only if the point AA' · BB' is interior.*

By 4·63 we now have:

7·54   *Two secants AA' and MN are conjugate lines if and only if* H(AA', MN).

The following two theorems will be used in Chapter 9, where we consider conics in Euclidean geometry:

7·55   *If a line is exterior to one of the conics that can be drawn through four given points, the quadrangular involution determined on the line is hyperbolic.*

Proof: This is obvious by considerations of continuity. For, since the quadrangular pencil of conics includes some for which the line is exterior as well as others for which it is a secant, it must include two for which it is a tangent. By 6·72, the points of contact are invariant points of the quadrangular involution.

Here is a better proof,* not using such considerations. By 3·14, applied to the four points on the given conic, two must separate the other two, say $PR||RS$. By 7·53, the intersection $T = PR \cdot QS$ is interior. Let the exterior line meet the sides of the quadrangle in A, A', B, B', C, C', as in Fig. 4·7B. Then $TB'||QS$ and $TB||PR$. Since

$$TB'QS \overset{P}{\barwedge} BB'CA', \quad TBPR \overset{Q}{\barwedge} B'BCA, \quad TB'QS \overset{R}{\barwedge} BB'AC,$$

we deduce
$$BB'||CA', \quad BB'||CA, \quad BB'||C'A.$$

Thus BB'/C and BB'/A are two supplementary segments, one containing both A and A', the other both C and C'. Hence A and A' do not separate C and C'. By 4·65, the involution $(AA')(CC')$ is hyperbolic.

7·56   *Two involutions, one or both elliptic, on the same line, always have a common pair of corresponding points.*

Proof: Transfer the two involutions to a conic, by the method of §7·4, and let their centres be O_i and O_j, as in Fig. 7·5B. Since at least one of these points is interior, their join O_iO_j is a secant, meeting the conic in M and N, say. Then MN is a common pair of the two involutions on the conic (cf. §4·6, Ex. 4).

---

* Kindly supplied by J.A. Jenkins.

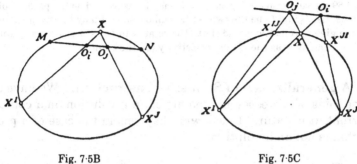

Fig. 7·5B   Fig. 7·5C

Incidentally, this shows that the product of the two involutions is hyperbolic, having M and N for invariant points. More generally, the product of *any* two involutions on a conic has for an axis the join of the two centres. For if the involutions transform X and X^I and X^J, respectively, as in Fig. 7·5C, their product

$$XX^I X^{JI} \, \overline{\wedge} \, X^{IJ} X^J X$$

has axis

$$(X^I X \cdot X^{JI} X^J)(XX^J \cdot X^I X^{IJ}) = O_i O_j.$$

EXERCISES

1. What happens to the involution with centre O when the conic degenerates into a line pair? Where then is the axis of the involution?

2. Use 7·53 to obtain a quick answer to Ex. 3 of § 4·7.

3. Given five points A, A', B, B', C, no three collinear, devise a linear construction for the point C' on the conic $AA'BB'C$ that is paired with C in the involution $(AA')(BB')$.

4. Adapt the method of § 7·4 so as to construct the points M and N of 3·62. These are, of course, the invariant points of the hyperbolic involution $(AB)(CD)$.

5. Show that each pair of an elliptic involution are harmonic conjugates with rspect to one other pair.*

6. Prove that two involutions on a conic commute if and only if their centres are conjugate.†

7. Let $AA_1 A'BB_1 B'$ be a hexagon inscribed in a conic. Show how Pascal's theorem leads to a new proof of 4·68.

---

* Holgate 1930, p. 213.
† Veblen and Young 1918, p. 227.

8. By 4·69 a given projectivity on a conic may be expressed as the product of two involutions. Let O_i and O_j be the two centres. Show that O_i may be any point on the axis of the given projectivity and that O_i is related to O_j by a projectivity (on the axis) of the same type as the given projectivity (on the conic).

7·6 A generalization of Steiner's construction. We have seen that the joins of corresponding points of an involution on a conic are concurrent. It is natural to ask what happens in the case of a projectivity that is not an involution.

7·61 *The joins of corresponding points of two projectively related ranges on a conic envelop a conic (provided that the projectivity is not an involution).*

*Proof:** Let A be a fixed noninvariant point on the conic and X a variable point on the conic, so that $AA'X \barwedge A'A''X'$, as in Fig. 7·6A. Then the axis of the projectivity is PQ, where

$$P = AX' \cdot XA', \quad Q = A'X' \cdot XA''.$$

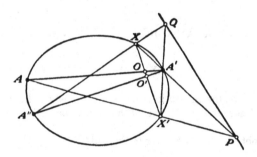

Fig. 7·6A

Let XX' meet the two fixed lines AA' and $A'A''$ in O and O'. From the quadrangle $AA'XX'$, O and P are conjugate points. From the quadrangle $A'A''XX'$, O' and Q are conjugate points. As X and X' vary on the conic, P and Q vary on the axis, O varies on AA', and O' on $A'A''$; thus

$$O \barwedge P \overset{A}{\barwedge} X' \overset{A'}{\barwedge} Q \barwedge O';$$

but O and O' cannot coincide (as A' is not an invariant point of the projectivity on the conic). Hence, by 6·58, the line OO' (or XX') envelops a conic.

---

* F.S. Macaulay 1906, p. 284. For the complex version of this theorem see Baker (1930, p. 52) or Coolidge (1945, p. 111). For another 'real' proof see Walker 1946.

EXERCISES

1. Two sides of a variable triangle inscribed in a conic pass through fixed non-conjugate points. Prove that the third side envelops a conic (cf. §6·4, Ex. 3).

2. Those tangents to one conic which cut another conic determine on the latter an ordered correspondence. Show that this is not, in general, a projectivity. (*Hint:* Five arbitrary pairs of points on the conic may be related by such a correspondence; but of course no more than three arbitrary pairs can be related by a projectivity.)

7·7 Trilinear polarity. We proceed to show how a triangle induces a correspondence between points not on its sides and lines not through its vertices. Although this is called a 'trilinear polarity', it is not really a polarity at all, for, as we shall see, the 'poles' of concurrent lines are not collinear points.

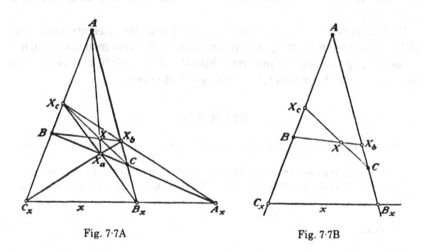

Fig. 7·7A   Fig. 7·7B

Given a triangle ABC and a point X not on any side, construct the six points

$$X_a = XA \cdot BC, \qquad X_b = XB \cdot CA, \qquad X_c = XC \cdot AB,$$

$$A_x = X_b X_c \cdot BC, \qquad B_x = X_c X_a \cdot CA, \qquad C_x = X_a X_b \cdot AB,$$

as in Fig. 7·7A. By Desargues's theorem for triangles ABC and $X_a X_b X_c$, the three points A_x, B_x, C_x lie on a line x. This is called the *trilinear polar* of X (with respect to ABC).

Since A_x, B_x, C_x are the harmonic conjugates of X_a, X_b, X_c on the respective sides,* we can easily reconstruct X, the *trilinear pole* of a given line x (not through any vertex).

---

* Poncelet (1865, vol. II, p. 34) used this property to *define* the trilinear polar.

Carrying over these definitions to the forbidden positions, we should say that the trilinear polar of a point on a side is that side itself, while the trilinear pole of a line through a vertex is that vertex itself; but the trilinear polar of a vertex and the trilinear pole of a side are indeterminate.

7·71 *The trilinear poles of a pencil of lines are the points of a conic circumscribing the triangle.*

Proof: Let x be a variable line through a fixed point O, not on a side of the given triangle ABC, and X its trilinear pole, as in Fig. 7·7B. From the involutions $(CC)(AA)$ and $(AA)(BB)$, we have

$$X_b \barwedge B_x \overset{O}{\barwedge} C_x \barwedge X_c;$$

and hence the locus of $X = BX_b \cdot CX_c$ is a conic through B, C, and similarly through A. (This is called the *polar conic* of O with respect to the triangle.)

If O, instead of being a point of general position, lies on a side, say BC, then we have $X_b \barwedge X_c$, and the conic degenerates into that side and a line through the opposite vertex. Finally, if O is taken at a vertex, the locus consists of the two sides through that vertex.

EXERCISES

1. Show that the line $A_1 B_1 C_1$ of Fig. 2·4C is the trilinear polar of the point S.

2. Prove that the dual of the above construction for x (given X) provides a construction for X (given x). (*Hint:* Show that the three lines AX, BB_x, CC_x are concurrent.)

3. Dualize 7·71.

4. If the trilinear polars of the vertices of a triangle (not ABC) are concurrent, prove that the trilinear poles of the sides are collinear.

5. Given a conic and an inscribed (or circumscribed) triangle, show that there is just one point whose polar with respect to the conic coincides with its polar with respect to the triangle. (*Hint:* Use §6·3, Ex. 3.)

6. Given an elliptic polarity and a self-polar triangle, show that there are just four points whose polars coincide with their trilinear polars. (*Hint:* Apply §7·5, Ex. 5, to the involution of conjugate points on two sides of the triangle.)

CHAPTER 8

Affine Geometry

Projective geometry, in marked contrast to ordinary Euclidean geometry, is not at all concerned with length or distance; it contains no criterion for telling whether two segments are 'congruent'. Affine geometry, however, takes us half-way back to the concept of distance: we are able to measure lengths along one line or on parallel lines and even to measure area, but we still cannot compare segments in different directions.

It is remarkable how many of the concepts and properties ordinarily considered in Euclidean geometry are still valid in the wider system of affine geometry. As we saw in § 1·2, such concepts and properties are just those which are invariant under parallel projection.

Klein treated affine geometry by means of coordinates, viz. oblique Cartesian coordinates with independent scales of measurement along the two axes. The details of the synthetic treatment were worked out by Veblen.

8·1 Parallelism.
Affine geometry can be derived from projective by singling out one line *o* and calling it the *line at infinity*, so as to be able to define parallelism. Any point on *o* is called a *point at infinity*, and two lines are said to be *parallel* if their intersection is such a point. Strictly, affine geometry is concerned only with 'ordinary' points and lines (not at infinity); hence we may say that the affine plane is derived from the projective plane by *removing* the line *o*. Thus two lines are parallel if they have no (ordinary) intersection. The theory could be built up in terms of ordinary points and lines alone, but we shall find it easier to make use of *o*, that is, to develop affine geometry from the projective point of view. All our theorems, however, will be stated in terms of ordinary points and lines.

It is an immediate consequence of our definition of parallelism that just one line can be drawn, through a given point, parallel to a given line (not passing through the point), and that all lines parallel to a given line are parallel to one another.

Four points are said to form a *parallelogram OACB* if *OA* is parallel to *BC* and *OB* to *AC*.

EXERCISES

1. Let *OAMA'* and *OBLB'* be two parallelograms having their sides at *O* along the same lines (*B* on *OA*, *B'* on *OA'*). Prove that the lines *AB'*, *BA'*, *LM* are concurrent or parallel. (*Hint:* Use 4·31)

2. Through a point *X* draw a line *AC* with *A* and *C* on two fixed parallel lines. Through a fixed point *O* (not on any of these lines) draw *OA*, and let the parallel line through *C* meet *OX* in *X'*. Prove that the position of *X'* is independent of the choice of the transversal *AC* and that the correspondence *X → X'* is a homology.* (*Hint:* Use the second part of Fig. 5·2B, with *A'* at infinity.)

8·2 Intermediacy. In §3·1 we considered a closed line on which the order of points is *cyclic*. The removal of a point at infinity changes this into an *open* line on which the order of points is *serial*, like the order of real numbers. In other words, separation is replaced by intermediacy: we say that *B* is *between* two given points *C* and *D* when *AB//CD*, where *A* is the point at infinity on the line *CD*. Axioms 3·11–3·16 lead to familiar properties of intermediacy:

8·21 *There exists a line containing three distinct points.*

8·22 *If B is between C and D, then B, C, D are three distinct collinear points.*

8·23 *Any point between C and D is also between D and C.*

8·24 *Of any three distinct collinear points, one is between the other two.*

8·25 *If B is between C and D, while C is between B and E, then B is between D and E* (as in Fig. 8·2A).

8·26 *Intermediacy is preserved by parallel projection* (Fig. 8·2B).

If we wished to build up affine geometry as an independent system instead of deriving it from projective geometry, we should take some of the above properties (along with certain statements about incidence) as a new set of axioms. Such a system is in some respects simpler than projective geometry; for *point* and *intermediacy* are the only primitive concepts needed: 'line' and 'incidence' can be defined in terms of them.

---

* La Hire, as quoted by Lehmer 1917, p. 110.

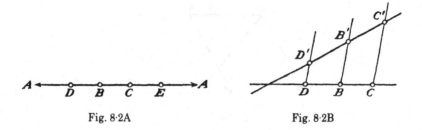

Fig. 8·2A   Fig. 8·2B

In affine geometry, any two points B and C determine a unique segment BC, namely, BC/A, where A is the point at infinity on the line BC. This simply consists of all points between B and C. Similarly, the *interior* of a triangle ABC is the region ABC/o, in the notation of §3·8.

Although an ordinary line is open, the line at infinity is still closed. Thus we cannot say that one of three points at infinity lies between the other two, but only that two of four points at infinity separate the other two. It is important to notice that a positive sense of rotation is determined at all (ordinary) points simultaneously by calling one of the two senses along o the positive sense.

EXERCISES

1. Deduce from 3·21 that the affine line is decomposed by any one of its points into two half-lines, or *rays*. If B lies between C and D, the two rays are naturally denoted by B/C and B/D. (Of course, B/C is the one that does *not* contain C.)

2. Develop the affine theory of sense, using a symbol $S(BC) \neq S(CB)$.

8·3 Congruence.* Two segments QQ' and RR' are said to be *congruent by translation* if $QQ'R'R$ is a parallelogram. We then write

$$QQ' \equiv RR'.$$

Theorem. 1·51 with O at infinity shows that the two relations $QQ' \equiv RR'$ and $RR' \equiv PP'$ imply $PP' \equiv QQ'$. We naturally extend this notion so as to allow PP' and QQ' to be collinear, as in Fig. 8·3A; we write $PP' \equiv QQ'$ whenever there is a segment RR' such that $PP'R'R$ and $QQ'R'R$ are parallelograms. (Desargues's theorem shows that the position of R is immaterial.) It follows from this extended definition that the relation 'congruent by translation' is reflexive ($PP' \equiv PP'$), symmetric (so that $PP' \equiv QQ'$ implies $QQ' \equiv PP'$), and transitive (so that $PP' \equiv QQ' \equiv RR'$ implies $PP' \equiv RR'$).

We proceed to show how the relation between congruent segments on one line may be expressed as a projectivity. In Fig. 4·4C we con-

---

* Veblen and Young 1918, p. 75.

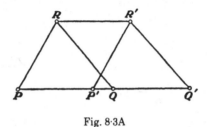

Fig. 8·3A

structed the parabolic projectivity

$$MMA \,\overline{\wedge}\, MMA'$$

by choosing points R and S on an arbitrary line through the invariant point M. If M is at infinity, we may take RS to be the line at infinity, so that AXX_0A_0 and $A'X'X_0A_0$ are parallelograms and $AX \equiv A'X'$. Conversely, if AX and $A'X'$ are congruent segments on the same line, we have parallelograms AXX_0A_0 and $A'X'X_0A_0$ from which we can reconstruct the figure for a parabolic projectivity. By 4·72, the relation

$$MMAX \,\overline{\wedge}\, MMA'X' \quad \text{implies} \quad MMAA' \,\overline{\wedge}\, MMXX';$$

that is

$$AX \equiv A'X' \quad \text{implies} \quad AA' \equiv XX'.$$

Hence:

8·31   *Two segments on the same line are congruent if and only if one is transformed into the other by a parabolic projectivity whose invariant point is at infinity; a variable segment XX' on a fixed line remains congruent to a fixed segment AA' if and only if X and X' are related by such a projectivity.*

If $AA' \equiv A'A''$, we call A' the *midpoint* of the segment AA''. By 4·43, this means that $H(MA', AA'')$, where M is the point at infinity on the line AA''. Hence, writing B for A'',

8·32   *The midpoint of a segment AB is the harmonic conjugate (with respect to A and B) of the point at infinity on the line AB.*[*]

EXERCISES

1. Consider what happens to Fig. 2·5A when CPQ is the line at infinity. Show that the diagonals of a parallelogram have the same midpoint. Deduce that the congruence $PP' \equiv QQ'$ (on one line or on parallel lines) is equivalent to the statement that PQ' and QP' have the same midpoint.

---

[*] von Staudt 1847, p. 204, § 338.

2. Prove that the medians of a triangle are concurrent.* (*Hint:* See § 7·7.)

3. Prove Varignon's theorem: The midpoints of the four sides of a simple quadrangle are the vertices of a parallelogram. Deduce that, if B is the midpoint of AC, and B' of $A'C'$, then the midpoints of AA', BB', CC' are collinear.†

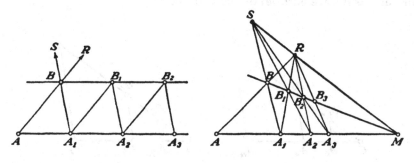

Fig. 8·4A

8·4 Distance. One a given line AA_1, let further points A_2, A_3, ... be taken so that

$$AA_1 \equiv A_1A_2 \equiv A_2A_3 \equiv \cdots.$$

Then we say that the *distance* AA_n is n times AA_1. Fig. 8·4A illustrates both the parabolic projectivity

$$AA_1A_2 \cdots M \overset{R}{\barwedge} BB_1B_2 \cdots M \overset{S}{\barwedge} A_1A_2A_3 \cdots M$$

and the harmonic relations $H(MA_1, AA_2)$, $H(MA_2, A_1A_3)$, $\cdots$. This gives us a rule for multiplying a given distance by any positive integer. Can we also construct a fraction of a given distance? Not immediately, but soon.

It is intuitively obvious that the segments A_1A_2, A_2A_3, ... cover the whole ray A_1/A (or A_1M/A), so that the point at infinity M is the limit of the sequence of A's and may appropriately be called A_∞. But the rigorous proof of this fact is postponed till § 10·2.

Suppose we have

$$OA_1 \equiv A_1A_2 \equiv A_2A_3 \equiv \cdots$$

on one line and

$$OB_1 \equiv B_1B_2 \equiv B_2B_3 \equiv \cdots$$

on another, as in Fig. 8·4B. Then, because the successive A's and B's

---

* von Staudt 1847, p. 204, § 338.
† Coxeter 1969, p. 47.

were constructed by taking harmonic conjugates, we have

$$OA_1 A_n A_\infty \barwedge OB_1 B_n B_\infty.$$

By 4·23, this projectivity is a perspectivity whose centre lies on the line at infinity $A_\infty B_\infty$. Hence $A_n B_n$ is parallel to $A_1 B_1$, and the lines $A_n B_n$ for various n's are parallel to one another.

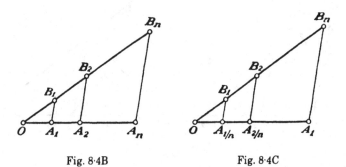

Fig. 8·4B   Fig. 8·4C

Now, to divide a given segment OA_1 into n equal parts, draw any other line through O, and take on it points $B_1, B_2, \ldots, B_n$ so that

$$OB_1 \equiv B_1 B_2 \equiv \cdots \equiv B_{n-1} B_n.$$

Join $A_1 B_n$ and draw parallel lines through the other B's to meet OA_1 in points $A_{1/n}, A_{2/n}, \ldots, A_{(n-1)/n}$, as in Fig. 8·4C. Then

$$OA_{1/n} \equiv A_{1/n} A_{2/n} \equiv \cdots \equiv A_{(n-1)/n} A_1,$$

and each of these segments is $1/n$ of IA_1.

   In this manner we can construct a segment $OA_{m/n}$ for any given fraction m/n, and it is not difficult to see that the order of such points $A_{m/n}$ agrees with the order of the rational numbers m/n. (We naturally use A_0 as an alternative symbol for O.) Considerations of continuity* then enable us to define A_x for any positive number x (whether rational or not).

   Negative numbers may be included by defining A_{-x} as the harmonic conjugate of A_x with respect to A_0 and A_∞, so that $A_{-x} A_0 \equiv A_0 A_x$; and then we write

$$A_0 A_{-x} = -A_0 A_x \quad \text{or} \quad A_0 A_x + A_0 A_{-x} = 0.$$

By 4·63, A_x and A_{-x} are a typical pair of the involution

$$(A_0 A_0)(A_\infty A_\infty).$$

---

   * The details are omitted because for this purpose Axiom 3·51 is quite unsuitable. The relevant treatment of continuity, based on the ideas of Weierstrass and Cantor, will be found in Chapter 10.

Instead of saying that the distance $A_0 A_x$ is x times $A_0 A_1$, we may say that the *ratio* $\dfrac{A_0 A_x}{A_0 A_1}$ is equal to the real number x. The above remarks provide a definition for the ratio of any two segments on one line or on parallel lines.* It is important, however, to realize that the ratio of two segments in any other relative position is essentially indeterminate. If OAB is a triangle, the symbol $\dfrac{OB}{OA}$ has no numerical value (in affine geometry); we cannot say whether OA or OB is 'longer'. What we can say about segments on intersecting lines is as follows:

8·41 *If A' is on OA and B' on OB, with $A'B'$ parallel to AB, then* $\dfrac{OA'}{OA} = \dfrac{OB'}{OB}$. *Conversely, if* $\dfrac{OA'}{OA} = \dfrac{OB'}{OB}$, $A'B'$ *must be parallel to AB.*

We are now ready to reconcile Desargues's treatment of involutions with von Staudt's. We have already considered the trivial case when the involution has an invariant point at infinity. In any other case the point at infinity is paired with an ordinary point C, called the *centre* of the involution (though it has no connexion with the O of 7·52).† When RS is the line at infinity, Fig. 4·7B reduces to Fig. 8·4D, with AQ parallel to BP and $A'P$ to $B'Q$. By 8·41,

$$\frac{CA}{CB} = \frac{CQ}{CP} = \frac{CB'}{CA'},$$

that is, $CA \times CA' = CB \times CB'$. Hence:

8·42 *If M is the point at infinity on the line AA', the involution $(AA')(CM)$ relates points X and X' such that the product $CX \times CX'$ is constant.*

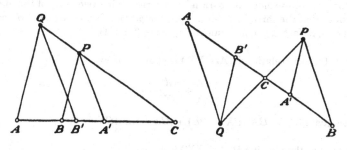

Fig. 8·4D

---

* Veblen and Young 1918, p. 85. The transitivity of the equality of ratios is a consequence of 2·25 in the special form 1·51.
† Young 1930, pp. 98–9.

Strictly, a product of distances is not defined. (Certainly we must not follow Euclid in speaking of $CX \times CX'$ as a 'rectangle'!) The above statement is easily expressed in terms of ratios: we mean that

$$\frac{CX}{A_0 A_1} \times \frac{CX'}{A_0 A_1}$$

is constant.

By 4·63, the centre of a hyperbolic involution is midway between the two invariant points. Hence:

8·43   *The relation* H(AB, XX') *is equivalent to*

$$CX \times CX' = CA^2,$$

where C is the midpoint of AB.

A 'trivial' involution, for which the point at infinity is invariant, has no centre in the above sense; but now another point takes over that role—the second invariant point A_0. A pair of corresponding points, being harmonic conjugates with respect to A_0 and A_∞, are equidistant from A_0 on opposite sides, i.e. the algebraic sum of their distances from A_0 is zero. More generally, taking an arbitrary origin O instead of A_0, the sum of their distances from O is constant.

EXERCISES

1. On a line that meets the sides of a triangle PQR in A, B, C, points A', B', C' are chosen so that the segments AA', BB', CC' all have the same midpoint. Prove that the lines PA', QB', RC' are concurrent. (*Hint:* See 4·71.)

2. Prove that the midpoints of two pairs of opposite sides of a quadrangle form a parallelogram (whose sides are parallel to the remaining two sides of the quadrangle). Deduce that the three lines joining the midpoints of opposite sides are concurrent (having, in fact, a common midpoint; see §8·3, Ex. 1).

3. Show that the relation H(AB, XX') is equivalent to

$$\frac{AX}{BX} + \frac{AX'}{BX'} = 0.$$

(*Hint:* Evaluate $(CX - CA)(CX' - CB) + (CX - CB)(CX' - CA)$.)

4. Show that the relation H(AB, XX') is equivalent to

$$\frac{1}{AX} + \frac{1}{AX'} = \frac{2}{AB},$$

so that AB is the *harmonic mean* between AX and AX'. (This is the origin of the name *harmonic conjugate*.)

8·5 Translation and dilatation.* Let o denote the line at infinity. An elation with axis o is called a *translation*, and a homology with axis o is called a *dilatation*. In particular, a harmonic homology with axis o (that is, an involutory dilatation) is a *half-turn*. In this case, by 8·32, the joins of pairs of corresponding points all have the same midpoint; consequently the half-turn is sometimes called *reflexion in a point* or *central inversion*. (We may think of it as the transformation that interchanges the numerals 6 and 9.)

In the first part of Fig. 5·2A (with o at infinity), $ABB'A$ is a parallelogram. Hence $AB \equiv A'B'$ if and only if there is a translation that takes AB to $A'B'$. By 5·22, the translation is determined when we are given the corresponding points A and A'. We naturally call it the *translation from A to A'*. The transitivity of the congruence relation is now revealed as a consequence of the fact that the product of the translation from P to Q and the translation from Q to R is the translation from P to R (see 5·25). The projectivity of 8·31 is induced by the translation in accordance with 5·23.

In this manner the notion of congruence can be extended from segments to figures of any kind: two figures are *congruent* if one can be derived from the other by a translation.

Similarly, two figures are said to be *homothetic* (or *similar and similarly situated*) if one can be derived from the other by a dilatation; e.g. two incongruent segments AB and $A'B'$ on parallel lines are homothetic from the centre $AA' \cdot BB'$. If instead the two incongruent segments are both on the same line whose point at infinity is M, they are related by the hyperbolic projectivity $MAB \overline{\wedge} MA'B'$. This is induced (according to 5·23) by a dilatation whose centre is the second invariant point of the projectivity, i.e. the centre of the involution $(AB')(BA')$ (see the preamble to 4·67). Hence:

8·51 *Any two incongruent segments, on one line or on parallel lines, are related by a dilatation.*

In particular, if $AB \equiv B'A'$, the dilatation from AB to $A'B'$ is a half-turn, and the centre is the common midpoint of AA' and BB'. This suggests the desirability of extending the meaning of congruent to include *congruent by a half-turn*, so that we can write

$$AB \equiv BA.$$

From now on we shall use congruent in this wider sense, which will cause no confusion since, by 5·28, any translation can be expressed as the product of two half-turns. Accordingly, instead of the distance AB,

---

* Veblen and Young 1918, pp. 92–5. A remarkably simple self-contained treatment of affine geometry (using 8·51 to *define* dilatation) has been given by Emil Artin (1940, pp. 15–20).

which may be positive or negative according to the sense, we consider the *length AB*, which is essentially positive. Then two segments are congruent if and only if they have the same length.

By 5·24, if the sides of one triangle are parallel to respective sides of another, the two triangles are either homothetic or congruent. They are congruent by translation if the joins of corresponding vertices are parallel, and congruent by a half-turn if the joins have a common mid-point.

EXERCISES

1. Show that the product of a translation and a half-turn is a half-turn.

2. Let C and C' be arbitrary points on the opposite sides AB and $A'B'$ of a parallelogram $ABB'A'$. Let the line $(BC' \cdot CB')(CA' \cdot AC')$ meet AA' in P and BB' in Q. Prove that $AP \equiv QB'$. (*Hint:* Use 4·31 and the symmetry of the parallelogram.)

8·6   Area.* In affine geometry we cannot compare lengths in different directions, but we can compare areas in any position, since the ratio of two areas is invariant under parallel projection (see § 1·2).

A region of the affine plane (not including any point at infinity) is called a *polygon* if it is entirely bounded by line segments. Clearly, any polygon can be dissected into a finite number of triangles. Two polygons are said to be *equivalent* if (1) they can be dissected into a finite number of pieces that are congruent in pairs, or (2) it is possible to annex to them one or more congruent pieces so that the completed polygons are equivalent in the first sense. In other words, two polygons are equivalent if they can be derived from each other by addition or subtraction of congruent pieces. By superposing two different dissections, we see that two polygons equivalent to the same polygon are equivalent to each other.

The parallelograms $OPRQ$ and $OPR'Q'$ of Fig. 8·6A are equivalent since the same trapezoid $OPR'Q$ is obtained by annexing triangle PRR' to the former or OQQ' to the latter. Hence:

8·61 *Two parallelograms are equivalent if they have one pair of opposite sides of the same length lying on the same pair of parallel lines.*

We could almost as easily prove this by taking the two parallelograms to have the same centre (instead of a common base), so that the dissection would be centrally symmetrical. Since a parallelogram can be dissected along a diagonal into two triangles that are congruent by

---

* This is essentially the treatment of Hilbert (1930, Chapter IV), simplified by admitting continuity, but generalized by avoiding the use of right angles. For a more rigorous treatment see Veblen and Young (1918, pp. 96–104).

a half-turn, it follows that two triangles are equivalent if they have congruent sides on one line and opposite vertices on a parallel line (i.e. equal bases and equal altitudes). The actual dissection for such a pair of triangles is illustrated in Fig. 8·6B.

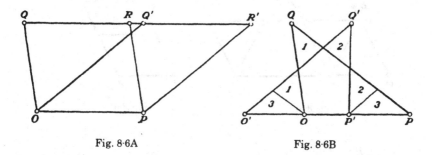

Fig. 8·6A   Fig. 8·6B

By annexing a further triangle OPQ' to each of two equivalent triangles with a common base PQ', as in Fig. 8·6C, we deduce the following:

8·62 *Two triangles OPQ and OP'Q', having a common angle at O, are equivalent if the lines PQ' and P'Q are parallel.*

By 'doubling' these triangles, as in Fig. 8·6D, we deduce:

8·63 *Two parallelograms OPRQ and OP'R'Q', having a common angle at O, are equivalent if the lines PQ' and P'Q are parallel.*

This suggests the propriety of selecting a certain parallelogram $OACB$ as the unit of measurement and defining the *area* of a parallelogram $OPRQ$, with P on OA and Q on OB, to be the number

$$\frac{OP}{OA} \times \frac{OQ}{OB}.$$

By 8·41 and 8·63, two such parallelograms $OPRQ$ and $OP'R'Q'$ are equi-

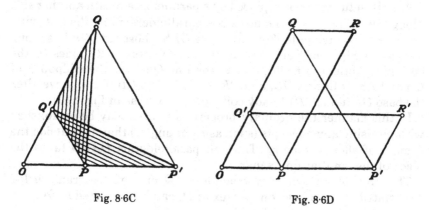

Fig. 8·6C   Fig. 8·6D

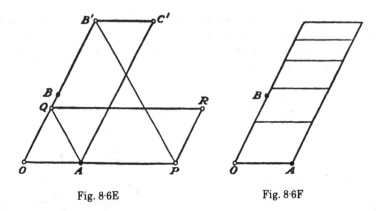

Fig. 8·6E   Fig. 8·6F

valent if $\dfrac{OP'}{OP} = \dfrac{OQ}{OQ'}$, i.e. if they have the same area. It follows that a parallelogram $OPRQ$, having two sides along the lines OA and OB, is equivalent to a parallelogram $OAC'B'$ where B' is constructed by drawing PB' parallel to AQ, as in Fig. 8·6E. The area is simply $\dfrac{OB'}{OB}$.

We now define the area of any polygon to be the area of an equivalent parallelogram $OAC'B'$. To see that the combined area of two or more juxtaposed polygons is the sum of their areas, we merely have to 'stack' the equivalent parallelograms, as in Fig. 8·6F.

When we wish to compute the area of a given polygon, we first dissect it into convenient pieces and then draw an equivalent parallelogram for each piece. A practical way to do this is to draw through each vertex of the polygon a line parallel to OA, thus dissecting the polygon into triangles, parallelograms, and trapezoids. We then dissect each trapezoid into two triangles by drawing a diagonal. Each piece is now a parallelogram or triangle having one side parallel to OA, and this can be translated to a position where the side proceeds from O along OA. This in turn may be replaced by a parallelogram with another side along OB, using 8·61 if the piece is a parallelogram and the following device if it is a triangle: For a triangle OPS whose side OP lies along OA, as in Fig. 8·6G, an equivalent parallelogram with sides in the desired directions is $OPRQ$, where the line QR joins the midpoints of OS and PS, Q lies on OB, and PR is parallel to OB. Finally, we alter the base OP (unless P already coincides with A) as in Fig. 8·6D.

In this manner the ordinary properties of area may be established without using any concepts (such as right angles) that are outside the domain of affine geometry. The unit parallelogram $OACB$ takes the place of the familiar unit square.

The area of an arbitrary triangle OPQ or parallelogram $OPRQ$ (translated so as to have one vertex at O) can be computed as follows:

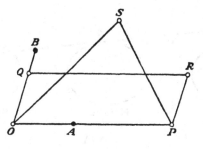

Fig. 8·6G

Let PQ meet OA in P', and OB in Q'; then

$$\frac{OPRQ}{OACB} = \frac{OPQ}{OAB} = \frac{OPQ}{OP'Q'} \frac{OP'Q'}{OAQ'} \frac{OAQ'}{OAB} = \frac{PQ}{P'Q'} \frac{OP'}{OA} \frac{OQ'}{OB}.$$

EXERCISES

1. In Fig. 8·6C (where $P'Q$ is parallel to PQ'), prove that

$$\frac{OP'Q}{OPQ'} = \left(\frac{OP'}{OP}\right)^2.$$

2. Give an affine proof for the following special case of Pappus's theorem: If alternate vertices of a hexagon lie on two intersecting lines, while two pairs of opposite sides are parallel, then the remaining sides are parallel.* *Hint:* In the notation of Fig. 4·3A with LN at infinity, we have equivalent triangles OAA', OBB', OCC', by 8·62. Alternatively, we can use 8·41 to obtain three equations such as

$$\frac{OA}{OB} = \frac{OB'}{OA'}.$$

8·7 Classification of conics.† The line at infinity, o, enables us to distinguish three types of conic: ellipse, parabola, and hyperbola. By definition, the conic is an *ellipse* if o is an exterior line (or e line), a *parabola* if o is a tangent, or a *hyperbola* if o is a secant (or h line). These are shown diagrammatically in Fig. 8·7A. By the dual of 6·56, a unique parabola can be drawn to touch the sides of a given quadrilateral.

The pole of o is the *centre*, O. Thus the centre of an ellipse is an interior point (E point), the centre of a parabola is its point of contact with o, and the centre of a hyperbola is an exterior point (H point). In the last case we can, of course, draw two tangents from the centre—

---

* Veblen and Young 1918, p. 103; cf. Pasch and Dehn 1926, p. 226.
† von Staudt 1847, pp. 138, 205, §§ 248, 341–3.

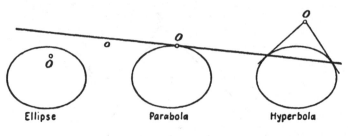

Fig. 8·7A

these are the *asymptotes* of the hyperbola (cf. § 1·3, Ex. 2). Their points of contact are on the line at infinity and decompose the hyperbola into two arcs, called the two *branches*.

Any line through O is a *diameter*. Since the centre of an ellipse is interior, all its diameters are secants. Since the centre of a parabola is at infinity, its diameters are parallel secants. Since the centre of a hyperbola is exterior, it has diameters of all kinds—the two asymptotes separate those which are secants from those which do not meet the hyperbola at all.

In the domain of ordinary points, a parabola has no centre. Accordingly the ellipse and hyperbola are called *central* conics.

When we apply 6·83 to the line at infinity, we find that the nine-point conic contains the midpoints of the sides of the quadrangle and the centres of all the conics of the quadrangular pencil. Hence:

8·71 *The midpoints of the six sides of a quadrangle and the three diagonal points all lie on a conic.*

EXERCISES

1. Show that a central conic is symmetrical about its centre. (Apply Ex. 4 of § 6·4 to the half-turn about O.)

2. If a diameter of a central conic is a secant, show that the tangents at its two 'ends' are parallel. On the other hand, a parabola has no parallel tangents.

3. Prove that the locus of centres of conics inscribed in a quadrilateral is a line through the midpoints of the three diagonals. (*Hint:* Use the dual of 5·81.) Deduce that this line is a diameter of the escribed parabola.

4. Show that the three lines joining the midpoints of opposite sides of a quadrangle are diameters of the nine-point conic (cf. § 6·8, Ex. 5 and § 8·4, Ex. 2).

5. Justify the construction of § 6·5, Ex. 6.

6. Prove that the centre of a conic cannot be interior to a self-polar triangle. (*Hint:* Use 6·21.)

7. Through a variable point X on a fixed line, a line c is drawn parallel to the polar of X with respect to a given ellipse. Prove that c envelops a parabola.*

8. Use 7·55 to define a *convex* quadrangle. What kind of quadrangle has just one circumscribed parabola?

8·8   Conjugate diameters.†   *A chord* of a conic is the segment joining two distinct points on the conic.

8·81   *The midpoints of parallel chords lie on a diameter.*

Proof: The parallel chords have a common point at infinity whose polar bisects them all, by 6·41 and 8·32. This polar, being conjugate to o, is a diameter conjugate to all the chords.

Hence, to construct a diameter, we merely have to join the midpoints of two parallel chords, as in Fig. 8·8A. The centre can be found as the point of intersection of two diameters.‡

By 5·53, conjugate diameters of a central conic are pairs of an involution. Since the invariant lines (if any) are asymptotes, this involution is elliptic or hyperbolic according as the conic is an ellipse or a hyperbola. (In fact, this is the origin of the names for the two types of involution, and thence, by analogy, the names *elliptic, parabolic, hyperbolic* for the three types of projectivity.) By 6·42, as La Hire observed:

8·82   *Any pair of conjugate diameters of a hyperbola are harmonic conjugates with respect to the asymptotes.*

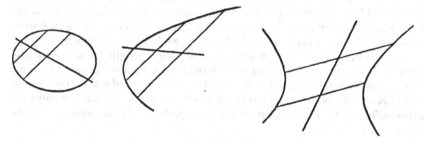

Fig. 8·8A

Two conjugate diameters form, with the line at infinity, a self-polar triangle, which has one e side and two h sides, as in Fig. 6·2C (second

---

* Chasles 1865, p. 138.

† Apollonius 1899, pp. 48–93, lib. I; Chasles 1865, pp. 114–17; Reye 1923, pp. 100–7.

‡ Apollonius 1891, pp. 265–7, lib. II, Prop. 44, 45.

part). Hence, in the case of an ellipse both diameters are secants, but in the case of a hyperbola one is a secant and the other an exterior line.

We shall make use of the next theorem in § 9·2, where we develop the theory of circles.

8·83 *If a parallelogram is inscribed in a conic, its two diagonals are diameters of the conic and its sides are parallel to a pair of conjugate diameters.*

Proof: The sides and diagonals of the parallelogram form a quadrangle whose diagonal triangle has two vertices at infinity, so one side of this triangle is *o*. By 6·43, its other two sides are conjugate diameters parallel to the sides of the parallelogram. (The conic, having the same centre as the parallelogram, obviously cannot be a parabola.)

The following theorem makes a neat companion for 8·81:

8·84 *The midpoints of chords that pass through a fixed point (not on an asymptote) lie on another conic.*

Proof: Let P be the point at infinity on the chord x through the fixed point A. This chord is bisected by the conjugate diameter p, and we have $x \barwedge P \barwedge p$. The lines x (through A) and p (through O) can never coincide unless OA is an asymptote. Hence, by 6·54, the point $x \cdot p$ lies on a certain conic through A and O.

Theorem 5·82 shows that we can find infinitely many conics having a given involution of conjugate points on a line a whose pole is a given point A. In particular, we can find infinitely many conics having a given centre and a given involution of conjugate points on o, i.e. having a given involution of conjugate diameters. If the involution is elliptic, the conics are ellipses, any two of which are related by a dilatation (see 6·84). If instead it is hyperbolic, the conics are hyperbolas having the same asymptotes (i.e. having double contact at infinity). They fall into two subsystems such that any two hyperbolas belonging to the same subsystem are related by a dilatation. Two belonging to different subsystems, however, are not so simply related,* for they are separated by their common asymptotes. To derive one from the other we need a homology whose centre is at infinity. In the special case when this is a *harmonic* homology (or *affine reflexion*) they are called *conjugate* hyperbolas. These remarks may be summarized as follows:

8·85 *A conic can be drawn through a given point so as to have two given pairs of concurrent lines as conjugate diameters. The conic will be an ellipse or a hyperbola according as the pairs do or do not separate each other. By varying the point we obtain, in the former case, a system*

---

* It is amusing to observe that Chasles 1865, pp. 246, 248) missed this little complication, although many properties of conjugate hyperbolas had been known since the time of Apollonius.

of homothetic ellipses all having the same involution of conjugate diame-
ters and, in the latter, two 'conjugate' systems of homothetic hyperbolas
all having the same asymptotes.

EXERCISES

1. Applying 8·81 to a hyperbola, investigate the nature of the diameter when the chords (a) join points on the same branch or (b) join a point on one branch to a point on the other. Observe that in the former case the diameter passes through the points of contact of two tangents parallel to the chords.

2. Prove that the diagonals of a parallelogram circumscribed about a central conic are conjugate diameters.

3. As a corollary of 8·83, any parallelogram inscribed in a conic is concentric with the conic. Deduce that the midpoints of the six sides of a quadrangle lie on a conic. (This provides an elementary proof for part of 8·71.)

4. Given two conjugate diameters a, b and a point P on the conic (but not on a or b), construct an inscribed parallelogram $PSQR$ whose sides are parallel to a and b.

5. Show that, when 8·84 is applied to a parabola, the locus is another parabola.*

6. Let aa' and bb' be two pairs of conjugate diameters of an ellipse. Prove that the sides of the parallelogram formed by the ends of a and b are parallel to those of the parallelogram formed by the ends of a' and b'. (*Hint:* Let m and n be the common harmonic conjugates of the pairs ab and $a'b'$. The involution of conjugate diameters interchanges these pairs. Being elliptic, it cannot leave m and n separately invariant and hence must interchange them.)

7. Use 7·53 to show that the number of (interior) intersections of the chords joining all pairs of n points, irregularly distributed on an ellipse, is $\binom{n}{4}$. (A.M. Gleason.)

8·9   Asymptotes.† The hyperbola provides a greater variety of special theorems than the other kinds of conic do, because of the existence of asymptotes. It is hoped that, after following the proofs of a few specimen theorems, the reader will feel prepared to deal with any affine properties of the hyperbola that may be proposed, either in the accompanying exercises or elsewhere.

8·91   *Any tangent to a hyperbola meets the two asymptotes in points equidistant from the point of contact.*

---

* Smith 1921, p. 137, no. 15.
† Apollonius 1891, pp. 198–231, lib. II, Prop. 3–21.

Proof: If *M* is the point of contact of the tangent *PQ*, as in Fig. 8·9A, the parallel diameter is conjugate to *OM*. By 8·82, these two diameters are harmonic conjugates with respect to the asymptotes *OP* and *OQ*. Thus *M* and the point at infinity on *PQ* are harmonic conjugates with respect to *P* and *Q*; i.e. *M* is the midpoint of *PQ*.

8·92 *If a chord AB of a hyperbola meets the asymptotes in P and Q, then PA ≡ BQ.*

Proof: The chord *AB* is bisected by the conjugate diameter, as we saw in the proof of 8·81 (see Fig. 8·9B), but 8·82 shows that this same diameter bisects the segment *PQ*. Thus *AB* and *PQ* have the same midpoint.

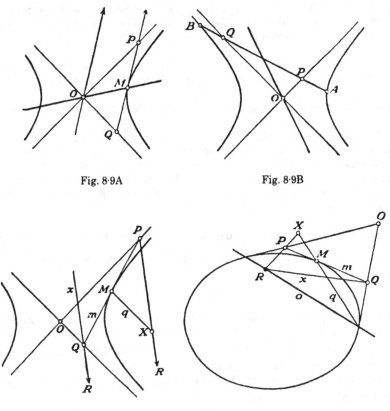

Fig. 8·9A   Fig. 8·9B

Fig. 8·9C

8·93 *If the tangent to a hyperbola at M meets the two asymptotes in P and Q, while MX is parallel to the latter asymptote, then the polar of X is the line through Q parallel to PX.*

Proof: Let *R* denote the point at infinity on *PX*, as in Fig. 8·9C. This point is conjugate to the vertex *O* of the triangle *OPQ*, which is circumscribed about the hyperbola. Hence, by 6·57, *PR* and *QR* are conjugate lines. Moreover, *MX*, joining the points of contact of the two tangents from *Q*, is the polar of *Q*. Thus *QR*, being conjugate to both *PR* and *MX*, is the polar of their point of intersection, *X*.

8·94   *A variable tangent to a hyperbola cuts off from the asymptotes a triangle of constant area.*

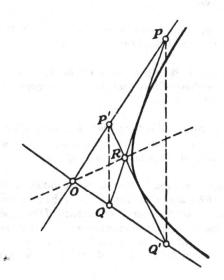

Fig. 8·9D

Proof: Let *PQ* and *P'Q'* be two tangents to the hyperbola, with *P* and *P'* on one asymptote and *Q* and *Q'* on the other, as in Fig. 8·9D. These two tangents form, with the two asymptotes, a quadrilateral *OPP'QQ'R*, circumscribed about the hyperbola. The three diagonal lines are *PQ'*, *P'Q*, and *OR*. The last is a diameter; therefore its pole, *PQ' · P'Q*, is at infinity, which means that *PQ'* and *P'Q* are parallel. By 8·62, triangles *OPQ* and *OP'Q'* have the same area.

8·95   Corollary. *If a variable line cuts off from two fixed lines a triangle of constant area, its envelope is a hyperbola, and the locus of the midpoint (of the segment intercepted) is the same hyperbola.* *

---

* Smith 1921, p. 203, no. 4.

EXERCISES

1. If through any point A a line APB is drawn parallel to an asymptote of a hyperbola, cutting the curve in P and the polar of A in B, show that P is the midpoint of AB.*

2. Prove that the midpoints of chords of a hyperbola that pass through a fixed point on an asymptote lie on a line parallel to the other asymptote. (*Hint:* Use 8·92.)

3. A variable segment has its ends on two fixed lines and passes through a fixed point. Prove that the locus of its midpoint is a hyperbola whose asymptotes are parallel to the given lines.†

4. If a tangent to a hyperbola meets the two asymptotes in P and Q, prove that any two parallel lines through P and Q are conjugate.

5. A parallelogram has its sides parallel to the asymptotes of a hyperbola, and one of its diagonals is a chord. Prove that the other diagonal passes through the centre of the hyperbola.*

6. Let PQ and $P'Q'$ be two parallel tangents to a hyperbola, with P and P' on one asymptote and Q and Q' on the other. Prove that PQ' and $P'Q$ are tangents to the conjugate hyperbola and that the parallelogram $PQ'P'Q$ has constant area.

7. When a quadrangle is convex, so that its nine-point conic with respect to o is a hyperbola, Ex. 5 of §6·8 shows that three pairs of the nine points are joined by diameters; therefore three points occur on one branch of the hyperbola while three take diametrically opposite positions on the other. Prove that the remaining three points (the diagonal points of the quadrangle) are all on one branch, i.e. that the nine points are distributed as $3 + 6$ between the two branches. (This is not easy‡.)

8·x   Affine transformations and the Erlangen programme. The translation and dilatation considered in §8·5 are special cases of an *affine collineation (or affinity)*, which is a collineation that leaves invariant the line at infinity (without necessarily leaving invariant each point at infinity). Applying 5·12 to two quadrilaterals both having the line at infinity for one side, we deduce

8·x1 *Any two triangles are related by a unique affine collineation.*‡
In particular, if two triangles have two common vertices while their remaining vertices lie on the same line parallel to their common side, they are related by a *shear*, which is an elation whose centre is at infinity although its axis is an ordinary line. By 5·21 with O at infinity, the shear with axis MN taking A to A' (so that MN and AA' are paral-

---

* Smith 1921, p. 204, nos. 11 and 12.

† Smith 1921, p. 203, no. 3.

‡ Neville 1960.

lel) has the following effect on other points X. Any X on MN is invariant. If X lies on AA', we have $XX' \equiv AA'$ (by translation). Any other X determines $C = AX \cdot MN$ and $X' = OX \cdot CA'$; that is, X' is the point where CA' meets the line through X parallel to the axis of the shear.

A shear transforms any triangle into another triangle having the same 'sense of description', i.e. the same sense for the points at infinity on their respective sides (so that, from this standpoint, PQR is equivalent to QRP or RPQ but not to RQP). This leads naturally to the concept of an *oriented* triangle, whose area is taken to be positive or negative according as its sense of description agrees or disagrees with that of a standard triangle OAB. The results of §8·6 enable us to assert that two oriented triangles have equal area (in sign as well as magnitude) if and only if they are related by an *equiaffine collineation*, which is a product of shears. Actually three shears always suffice, one to relate each pair of corresponding vertices in turn (in a suitable order).

In the celebrated 'Erlangen programme' of 1872, Klein proposed to classify geometries according to the groups of transformations under which their propositions remain true. In most cases the group has a subgroup that preserves the principal concepts belonging to the geometry. The meaning of these ideas becomes clear when we consider the following three instances:

Projective geometry consists of propositions that are preserved by the group of collineations and correlations. The subgroup of collineations preserves the class of points and the class of lines.

Affine geometry consists of propositions that are preserved by the group of affine collineations. The subgroup of equiaffine collineations preserves the measure of area.

Euclidean geometry (which is the subject of our next chapter) consists of propositions that are preserved by the group of similarities. The subgroup of congruent transformations preserves the measure of length.

EXERCISE

Express a translation as the product of two shears and a half-turn as the product of three shears.

CHAPTER 9

Euclidean Geometry

The time has come for us to fulfil the promise of §1·8, that we should return to ordinary geometry from a new point of view. We shall see how von Staudt's idea of choosing an elliptic involution on the line at infinity of the affine plane enables us to define perpendicularity and congruence, so that distances can be compared in any direction. Many problems of Euclidean geometry are most easily solved by the projective approach, but at this stage we are free to use either the new method or the old, whichever is found more convenient at the moment.

9·1 Perpendicularity. We have seen that affine geometry can be derived from real projective geometry by singling out for special treatment a line 'at infinity', which enables us to say when two other lines are parallel. Similarly, we shall find that Euclidean geometry can be derived from affine geometry by singling out an elliptic involution on that special line, to serve as the *absolute* involution, which enables us to say when two lines are perpendicular. From the standpoint of projective geometry, all elliptic involutions are exactly alike; but as soon as we have specialized one such involution, we can say (as a definition) that two lines shall be called *perpendicular* if their points at infinity are a pair of the absolute involution.

To see that this agrees with our usual ideas about perpendicularity, we merely have to observe that the correspondence between perpendicular lines is symmetric and preserves harmonic sets, so that it is an involution—elliptic because no line is perpendicular to itself.

Many properties of perpendicularity are immediate consequences of this definition; e.g. two perpendiculars to one line are parallel. The perpendicular line through the midpoint of a given segment is called the *right bisector* of the segment. A parallelogram that has two perpendicular sides is called a *rectangle*. A triangle that has two perpendicular sides is called a *right triangle*; any other kind of triangle is said to be *oblique*. The *altitudes* of a triangle are defined to be the perpendiculars from the vertices to the respectively opposite sides. The 'feet' of the altitudes are said to form the *orthic* (or *pedal*) triangle.

9·11   *The three altitudes of a triangle are concurrent.*

*Proof:** For a right triangle this is trivial, so let us assume the given triangle *PQR* to be oblique, as in Fig. 9·1A. Let the altitudes from *P* and *Q* intersect in *S*. Then the quadrangle *PQRS* determines on the line at infinity a quadrangular set of points, two of whose pairs belong to the absolute involution. Hence, by 4·71, the third pair likewise belongs to this involution, and *RS* must be the third altitude of the triangle.

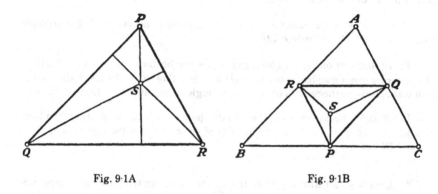

Fig. 9·1A   Fig. 9·1B

The point *S* is called the *orthocentre* of triangle *PQR*. (In the case of a right triangle, with *PR* perpendicular to *QR*, the orthocentre coincides with *R*.)

9·12   Corollary. *If S is the orthocentre of an oblique triangle PQR, then P is the orthocentre of QRS, and so on. The four triangles all have the same pedal triangle, namely, the diagonal triangle of the quadrangle PQRS.*

We define a *rectangular* hyperbola as one whose asymptotes are perpendicular (or whose points at infinity are a pair of the absolute involution). Since the absolute involution is elliptic, 7·55 shows that a conic through the vertices and orthocentre of an oblique triangle cannot be an ellipse. It cannot be a parabola, since then its centre would be an invariant point of the Desargues involution on the line at infinity. Hence, by 6·72:

9·13   *Every conic through the vertices and orthocentre of an oblique triangle is a rectangular hyperbola.*†

---

* von Staudt 1847, p. 205, § 339.
† Baker 1930, p. 83.

If P, Q, R are the midpoints of the sides of a triangle ABC, as in Fig.
9·1B, the right bisectors of those sides are the altitudes of the medial
triangle PQR. Hence:

9·14 *The right bisectors of the three sides of a triangle are concurrent.*
The point of concurrence (which is the orthocentre of PQR) is called
the *circumcentre* of ABC.

EXERCISES

1. Prove that those chords of a conic which subtend a right angle at any fixed
point on the conic are concurrent.*

2. Show that every rectangular hyperbola through the vertices of a triangle
passes also through the orthocentre.

3. Prove that just one rectangular hyperbola can be drawn through the vertices
of a quadrangle not consisting of a triangle and its orthocentre. (By Ex. 2, this passes
also through the orthocentres of the four triangles determined by these vertices.†

4. Let A and B be fixed points, C a variable point whose locus is either (i) a line,
or (ii) an ellipse. Find the locus of the point of intersection of the right bisectors of
AC and BC. (V. Hlavatý.)

9·2 Circles. In affine geometry there is no distinction of shape between one ellipse and another; although we can say which has the
greater area, we cannot say which has the greater 'eccentricity'. The
absolute involution enables us to make this further distinction and, in
particular, to define a circle.

Definition: A *circle* is a conic for which the involution of conjugate
diameters coincides with the involution of perpendicular diameters.
By applying 8·85 to two pairs of perpendicular lines through one
point, we verify that such a conic exists. In fact:

9·21 *Just one circle may be described with any centre to pass through*
any given point.
Thus we have proved Euclid's third postulate. It follows that a unique circle can be drawn with any given segment for a diameter. When
a circle is drawn with centre O to pass through a point A, the segment
OA is called a *radius*.
The diameter conjugate to a given chord is its right bisector. Conversely, any point O on the right bisector of a given segment AB is the

---

* von Staudt 1847, p. 206, §344. The point of concurrence is known as the
Frégier point. It lies on the normal to the conic at the given point.
† See Holgate 1930, p. 207.

centre of a circle through A and B; for the circle with radius OA must pass through B also, since the line AB is conjugate to the right bisector. Hence the circumcentre of a triangle ABC (9·14) is the centre of a circle through the three vertices:

9·22   *Any three non-collinear points lie on a unique circle.*
This is called the *circumcircle* of the triangle.

Since any two conjugate diameters are perpendicular,

9·23   *The polar of a point with respect to a circle is perpendicular to the diameter through the point.*
In particular:

9·24   *A tangent to a circle is perpendicular to the diameter through its point of contact.*

Since the involution of perpendicular diameters is elliptic, *a circle is an ellipse.* Ellipses that are not circles may conveniently be called *eccentric* ellipses. Since an involution is determined by two of its pairs,

9·25   *If a conic has two distinct pairs of perpendicular conjugate diameters, it must be a circle.*

By 8·83, any parallelogram inscribed in a circle is a rectangle whose centre is the centre of the circle. Hence:

9·26   *The lines joining the ends of a diameter to any other point on the circle are perpendicular.*
In other words, 'the angle in a semicircle is a right angle' (Euclid III.31). Conversely,

9·27   *The locus of the point of intersection of perpendicular lines through two fixed points is a circle.*
Proof: Let M and N be the points at infinity on the two perpendicular lines x and y. Since $x \barwedge M \barwedge N \barwedge y$, the locus is a conic (6·54) that has infinitely many inscribed rectangles with two fixed opposite vertices. By 8·83 and 9·25, any two of these rectangles suffice to make the conic a circle.

By 8·85, concentric circles are homothetic, but we can say far more:

9·28   *Any two circles are either congruent or homothetic.*
Proof: We remark first that a translation or dilatation, being a collineation that leaves invariant every point at infinity, transforms perpendicular lines into perpendicular lines and circles into circles. Let the line joining the centres of two given circles determine respective diameters AB and CD, and let M be the point at infinity on this line. Applying 8·51, we see that if $AB \not\equiv CD$ (as in Fig. 9·2A), one of the

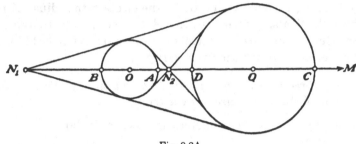

Fig. 9·2A

circles is transformed into the other by a dilatation from the centre of either of the involutions

$$(AD)(BC), \quad (AC)(BD).$$

(These points, N_1 and N_2, are called the *centres of similitude* of the two circles.) On the other hand, if $AB \equiv CD$, one of the dilatations has to be replaced by a translation, while the other becomes a half-turn.

Some of the above ideas suggest an elementary proof for the famous theorem of the nine-point circle:

9·29 *If S is the orthocentre of a triangle PQR, then the midpoints of the six segments QR, RP, PQ, PS, QS, RS and the feet of the three altitudes all lie on a circle.*

Proof:* Let L, M, N, L', M', N' be the six midpoints and PA, QB, RC the three altitudes, as in Fig. 9·2B. By 8·41, both NM and $M'N'$ are parallel to QR, and both MN' and NM' parallel to PS. Thus $MNM'N'$

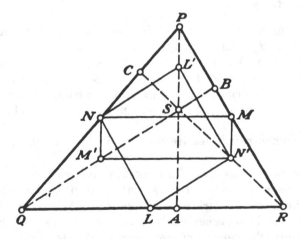

Fig. 9·2B

---

* This proof, due to V.E. Dietrich, resembles that of Durell 1931, p. 27.

is a rectangle. Similarly $NLN'L'$ is a rectangle. Hence LL', MM', NN' are three diameters of a circle. Since LA and $L'A$ are perpendicular, this circle passes through A; similarly, it passes through B and C.

It is interesting to observe that this is a special case of the nine-point conic (8·71). Thus the centres of the rectangular hyperbolas of 9·13 all lie on the nine-point circle.

A certain dilatation from the orthocentre S will transform the midpoints of PS, QS, RS into the vertices P, Q, R themselves and will consequently transform the nine-point circle of triangle PQR (which is the circumcircle of ABC) into the circumcircle of PQR. In other words, the nine-point circle is the locus of a point midway between the orthocentre and a point that runs round the circumcircle. It follows that the nine-point centre is midway between the orthocentre and circumcentre. The line on which these three points lie is called the *Euler line* of triangle PQR.

Theorems 9·26 and 9·27 show that the above definition for a circle is equivalent to Euclid's. Hence we can now reconcile von Staudt's definition for a conic with the classical one. For, any plane section of a circular cone is a central projection of a circle (as described in § 1·3).

EXERCISES

1. If a triangle is self-polar with respect to a circle, show that its orthocentre is the centre of the circle. (*Hint:* Apply 9·23 to each vertex of the triangle.)

2. Prove that the *centroid* (§ 8·3, Ex. 2) lies on the Euler line.*

3. Prove that the centres of similitude are harmonic conjugates with respect to the centres of the two circles. (*Hint:* In the notation of Fig. 9·2A we have

$$N_1 A \times N_1 D = N_1 B \times N_1 C \quad \text{and} \quad N_2 A \times N_2 C = N_2 B \times N_2 D;$$

therefore $N_1 O/N_1 Q = OA/QC = -N_2 O/N_2 Q$.)

4. Let CA and CB be two tangents to a circle with centre O. Join A, B, C to any point S on the circle by lines cutting the diameter perpendicular to OS in A', B', C'. Prove that C' is the midpoint of $A'B'$. (*Hint:* SA and SB are harmonic conjugates with respect to SC and the tangent at S.) This could have been made into an affine theorem by changing the words *circle* and *perpendicular* into *central conic* and *conjugate*, but the Euclidean theorem has an interesting application to solid geometry; for it shows that, when a small circle AB on a sphere is projected stereographically from S, the centre C' of the projected circle comes from the vertex C of the corresponding enveloping cone.†

9·3 Axes of a conic. An *axis* of a conic is defined as a diameter which is perpendicular to the chords it bisects (i.e. it is an axis of

---

* Robinson 1940, p. 48.
† Donnay 1945, p. 10.

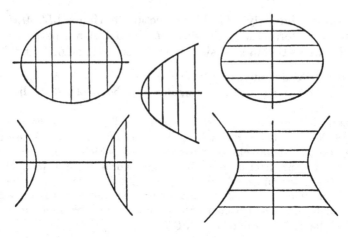

Fig. 9·3A

symmetry; see Fig. 9·3A). In the case of a parabola, all diameters are parallel, and there is just one system of chords perpendicular to them. Hence:

9·31   *A parabola has just one axis.*

This can be constructed by joining the midpoints of two chords drawn perpendicular to any diameter* (and the diameter can be constructed by joining the midpoints of any two parallel chords).

In the case of a central conic (ellipse or hyperbola) we consider two involutions of diameters:

(1) Conjugate diameters (an elliptic or hyperbolic involution).
(2) Perpendicular diameters (always an elliptic involution).

For a circle, these two involutions coincide. In every other case we can find (by the dual of 7·56) just one common corresponding pair. This is a pair of perpendicular conjugate diameters, i.e. a pair of axes. Hence:

9·32   *A central conic, other than a circle, has just two axes.*

A circle, however, has infinitely many axes—every diameter is an axis.

The axes of an ellipse (like all its diameters) are secants. Likewise the axis of a parabola is a secant (though one of the intersections is at infinity). In the case of a hyperbola, one axis is a secant while the other is an exterior line; these are called the *transverse* axis and *conjugate* axis, respectively.

---

* Apollonius 1891, pp. 267–71, lib. II, prop. 46–8.

When an axis is a secant, the points where it cuts the conic are called *vertices*. Thus a parabola has one vertex (unless we agree to admit another at infinity), a hyperbola has two vertices, and an eccentric ellipse has four.* A circle has infinitely many—every point on the circle is a vertex.

EXERCISES

1. Show that the circle on the transverse axis of a hyperbola as diameter has double contact with the hyperbola and that the circle on either axis of an eccentric ellipse as diameter has double contact with the ellipse.

2. Consider a conic and a point P not on it nor at its centre (if it has a centre). Let y be the line through P perpendicular to the chords bisected by a variable diameter x. Prove that the locus of $x \cdot y$ is a rectangular hyperbola through P. (This is known as Apollonius's hyperbola. It passes through the feet of the normals that can be drawn from P to the given conic.)

9·4 Congruent segments. In affine geometry we were able to define congruent segments on one line or on parallel lines. In Euclidean geometry we can define congruent segments in different directions, as follows:

Segments OA and OB are said to be *congruent by rotation* if A and B lie on a circle with centre O. Then we write

$$OA \equiv OB.$$

(Radii of a circle are congruent.) If a translation takes OA and OB to two radii $O'A'$ and $O'B'$ of another circle (so that $OA \equiv O'A'$ and $OB \equiv O'B'$ by translation), we simply say that OA and $O'B'$ are *congruent*, writing $OA \equiv O'B'$. This relation is clearly reflexive, symmetric, and transitive. The method of §8·4 now enables us to define the *length* of any segment in terms of a standard segment as unit. In particular, if C is exterior to the circle with radius OA, we say that OC is *greater* (or longer) than OA, writing

$$OC > OA.$$

If two axes of an ellipse are congruent, their four ends (forming a square) lie on a circle that touches the ellipse at all four places and therefore coincides with it entirely. Hence *an ellipse having equal axes is a circle*. Apart from this case, one axis of the ellipse must be longer than the other. The longer and shorter are called the *major* and *minor* axes.

Conjugate points on a diameter of a circle are said to be *inverse* with

---

* von Staudt 1847, p. 206, § 342.

respect to the circle. By 8·42, two inverse points have a constant product of distances from the centre. Since any point on the circle is its own
inverse, this product is equal to the square of the radius. In virtue of
9·23, we may express this result as follows:

9·41 *For a circle of radius* ρ, *the product of the central distances of a
point and its polar is equal to* ρ².

EXERCISE

Show that any affine property of a parallelogram can be deduced from the corresponding property of a square. Hence (or otherwise) prove the following affine
theorem:

Of all the ellipses circumscribing a given parallelogram, the one for which the
diagonals are *conjugate* diameters has the smallest area. (*Hint:* We know from elementary analytic geometry that the ellipse

$$\frac{x^2}{a^2} + \frac{y^2}{b^2} = 1,$$

of area πab, is circumscribed about the square (±c, ±c) if

$$\frac{1}{c^2} = \frac{1}{a^2} + \frac{1}{b^2} = \frac{2}{ab} + \left(\frac{1}{a} - \frac{1}{b}\right)^2 .)$$

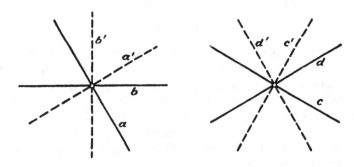

Fig. 9·5A

9·5 Congruent angles. Let *a* and *b* be two intersecting lines,
while *a'* and *b'* are the respective perpendiculars through their common point. Let *c, d, c', d'* be another such set of concurrent lines. The
ordered pair of lines ⟨*ab*⟩ is called an *angle*. We say that the two
angles ⟨*ab*⟩ and ⟨*cd*⟩ are *congruent* if the following two conditions are
satisfied:

(i) *aba'b'* ⊼ *cdc'd'*,

(ii) S(*aba'*) = S(*cdc'*)

(see Fig. 9·5A). We then write

$$\langle ab \rangle \equiv \langle cd \rangle.$$

This is clearly an equivalence relation, like the congruence of segments. It holds if a and b are respectively parallel to c and d (in which case $aba'b' \; \overset{o}{\barwedge} \; cdc'd'$). Moreover, $\langle ab \rangle \equiv \langle a'b' \rangle$.

Angles $\langle ab \rangle$ and $\langle ba \rangle$ are said to be *supplementary*. The angle $\langle aa' \rangle$ is called a *right* angle, and we regard all right angles as being congruent.

9·51   *The only angle congruent to its supplement is a right angle.*

Proof: Given $\langle ab \rangle \equiv \langle ba \rangle$, we wish to show that $b = a'$ (and consequently $b' = a$). If this were not so, we should have four distinct lines a, b, a', b' such that

$$S(aba') = S(bab') \neq S(abb'),$$

whence $ab // a'b'$, but since the involution of perpendicular lines $(aa')(bb')$ is elliptic, we have $aa' // bb'$. This provides the desired contradiction.

If three concurrent lines a, b, m have the property $\langle am \rangle \equiv \langle mb \rangle$, we say that m *bisects* the angle $\langle ab \rangle$.

9·52   *If a line m bisects $\langle ab \rangle$, then the perpendicular line n does likewise, and $H(mn, ab)$.*

Proof: We have $ama'n \; \barwedge \; mbnb' \; \barwedge \; b'nbm$, by the dual of 2·71. Therefore mn is a pair of the involution $(ab')(a'b)$. It is also a pair of the orthogonal involution $(aa')(bb')$. Hence m and n are the invariant lines of the product of these two involutions, which is $(ab)(a'b')$. This proves that $H(mn, ab)$. Moreover,

$$ana'm \; \barwedge \; mb'nb \; \barwedge \; nbmb',$$

and, reversing the sense

$$S(ama') = S(mbn),$$

we have $S(ana') = S(nbm)$; therefore $\langle an \rangle \equiv \langle nb \rangle$.

It follows similarly that the relation $\langle am \rangle \equiv \langle cm \rangle$ for concurrent lines a, m, c implies $a = c$; therefore the same relation for nonconcurrent lines implies that a and c are parallel.

The converse of 9·52 was discovered by Desargues.*

9·53   *If $H(mn, ab)$ and m is perpendicular to n, then m and n bisect the angles between a and b.*

---

* See Coolidge 1945, p. 29.

Proof: Applying the orthogonal involution to the given harmonic set, we have H(nm, $a'b'$). Thus ab and $a'b'$ are two pairs of the involution $(mn)(nn)$, and $ama'n \barwedge bmb'n \barwedge mbnb'$. We could not have $\langle am \rangle \equiv \langle bm \rangle$ without a and b coinciding. Hence $\langle am \rangle \equiv \langle mb \rangle$.

9·54 Corollary. *Angles that have a fixed pair of bisectors form an involution of line pairs.*

Another important property of angles is the following:

9·55 *If two angles $\langle ad \rangle$ and $\langle bc \rangle$ have the same bisectors, then*

$$\langle ab \rangle \equiv \langle cd \rangle.$$

Proof: If m and n are the bisectors, the four pairs ad, bc, $a'd'$, $b'c'$ belong to the involution $(mm)(nn)$. Therefore

$$aba'b' \barwedge dcd'c' \barwedge cdc'd'.$$

Moreover, since S(aba') $\neq$ S(dcd'), we have

$$S(aba') = S(cdd') = S(cdc').$$

In dealing with angles at different points, it is convenient to use another notation. If a is AO and b is OB (so that A and B are arbitrary points on the lines a and b, which intersect in O), we write $\langle AOB \rangle$ instead of $\langle ab \rangle$.

It should be emphasized that our definition of angle, while agreeing with Johnson's directed angle and Picken's cross (§1·6), differs from the customary definition, where the angle AOB would be changed into its supplement by moving B along the line b to the other side of the vertex O, and where BOA would be considered equal to either AOB or its negative (certainly not its supplement, as in the new treatment). The present convention has some definite advantages, chiefly in avoiding the separate consideration of various cases, as in the following theorem, where we should ordinarily have to make different statements according as S(ACB) agrees or disagrees with S(ADB):

9·56 *For any four points A, B, C, D on a circle,*

$$\langle ACB \rangle \equiv \langle ADB \rangle.$$

Proof: The angle $\langle ACB \rangle$ is measured by means of the four lines joining C to the vertices of the rectangle $ABA'B'$ whose diagonals are diameters AA' and BB' of the circle, as in Fig. 9·5B. Joining D, instead of C, to these same vertices, we obtain a related set of four lines and thence a congruent angle $\langle ADB \rangle$. To verify that the sense of the lines joining C to A, B, A' is independent of the position of C on the circle, we apply 7·55 to the quadrangle $ABA'C$ and the line at infinity (which is exterior since the circle is an ellipse). Since the quadrangular involution is hyperbolic, the sense of the points at infinity on the sides CA,

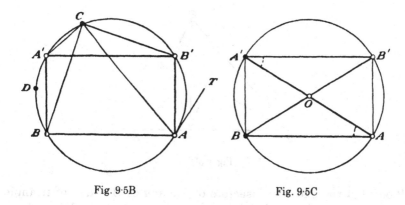

Fig. 9·5B   Fig. 9·5C

CB, CA' is opposite to that of the points at infinity on the remaining sides BA', $A'A$, AB, which are independent of C.

Conversely, the circle ABC could be described as the locus of a point X such that $\langle ACB \rangle \equiv \langle AXB \rangle$.

Another famous theorem is Euclid I.5 (*pons asinorum*):

9·57 *If $OA \equiv OB$, then $\langle OAB \rangle \equiv \langle ABO \rangle$.*
Proof: Let AOA' and BOB' be diameters of the circle with radii OA and OB, as in Fig. 9·5C. Applying the half-turn about O and using 9·56, we have

$$\langle OAB \rangle \equiv \langle OA'B' \rangle = \langle AA'B' \rangle \equiv \langle ABB' \rangle = \langle ABO \rangle.*$$

Theorem 9·53 helps us to obtain the following less trivial result:

9·58 *The altitudes and sides of an oblique triangle bisect the angles of its orthic triangle.*
Proof: Let ABC be the orthic triangle of PQR, as in Fig. 9·5D. Since ABC is the diagonal triangle of the quadrangle $PQRS$, the four lines drawn through A (or B or C) are a harmonic set, by 2·52.

It follows that each of the four points P, Q, R, S lies in a different one of the four regions determined in the projective plane by the lines BC, CA, AB. If S is the one lying in the finite region ABC/o, we say that AS and AQ are the *internal* and *external* bisectors of the angle A of triangle ABC.

9·59 *The bisectors of the angles of any triangle concur in sets of three to form a quadrangle.†*

---

* Here we are using the sign $\equiv$ for congruent angles and $=$ for identical angles. In § 1·6, where angles were regarded as magnitudes, this distinction was unnecessary.
† von Staudt 1847, p. 209, § 347.

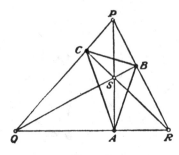

Fig. 9·5D

Proof: Let the external bisectors of the angles B and C of triangle ABC meet in P. We merely have to construct, as in 2·42, the quadrangle $PQRS$ whose diagonal triangle is ABC.

Of the four points P, Q, R, S, the one in ABC/o is called the *incentre* of triangle ABC, while the other three are the *excentres*.

EXERCISES

1. If AT is the tangent at A to the circle ABC, prove that

$$\langle ACB \rangle \equiv \langle TAB \rangle$$

(Fig. 9·5B).

2. Show that the centre of a circle lies on a bisector of the angle formed by two intersecting tangents.

3. Prove that the incentre and excentres are the centres of four circles touching the sides of the triangle.

4. Prove that the axes of a hyperbola bisect the angles between the asymptotes.

5. Prove that the asymptotes of a rectangular hyperbola bisect the angles between any pair of conjugate diameters.*

9·6 Congruent transformations. We define the *reflexion* in a line m to be the harmonic homology whose axis is m while its centre is at infinity in the perpendicular direction. Thus any two points A and B are interchanged by the reflexion in the right bisector of the segment AB, and any two intersecting lines a and b are interchanged by the reflexion in either of the bisectors of the angle $\langle ab \rangle$. Moreover, if OA and OB are congruent segments on a and b, a definite one of the angle bisectors will serve to reflect A into B. By the dual of 5·27, the product of two harmonic homologies having the same centre is an elation.

---

* Chasles 1865, p. 117.

Hence:

9·61 *The product of reflexions in two parallel lines is a translation.*

We define a *rotation* about a point O to be the product of reflexions in two lines through O. One instance has already been considered: By 5·32 and §8·5,

9·62 *The product of reflexions in two perpendicular lines is a half-turn.*

We define a *congruent transformation* to be a collineation that preserves length (and consequently preserves the line at infinity and the absolute involution). As instances we have a reflexion, a translation, a rotation and, more generally, the product of any number of reflexions.

By 5·12, a collineation is determined by its effect on any quadrilateral. Hence a congruent transformation is determined by its effect on a triangle (which provides a quadrilateral when we add the line at infinity). An alternative name for congruent transformation is *isometry*.

9·63 *A congruent transformation that leaves two points invariant is either the identity or a reflexion.*

Proof: Let P and Q be the invariant points. Then the transformation relates two triangles PQR and PQR'. If $R = R'$, we have the identity (by 5·11). Otherwise the altitudes from R and R' must have the same 'foot' C, and $RC \equiv R'C$; thus the transformation is a harmonic homology with axis PQ, in fact, the reflexion in PQ.

9·64 *Any congruent transformation that leaves just one point invariant is a rotation.*

Proof: Let the given transformation take a triangle PQR to $PQ'R'$. Then PQ is transformed into PQ' by the reflexion in a definite one of the bisectors of $\langle QPQ' \rangle$, say m. We now have two triangles $PQ'R_1$ and $PQ'R'$, as in Fig. 9·6A. These must be distinct, since otherwise the reflexion in m would suffice and every point on m would be invariant.

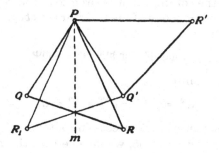

Fig. 9·6A

Hence the given transformation is the product of reflexions in m and PQ'.

Our next theorem is reminiscent of 4·22:

9·65 *Any congruent transformation can be constructed as a product of reflexions, the number of which can be reduced to three.*

Proof: If PQR is transformed into an entirely distinct triangle $P'Q'R'$, we begin by reflecting in the right bisector of PP' and then use one or two further reflexions as above.

In the projective plane, the senses of two pencils may be compared only if the pencils have the same centre; but in the affine plane the translation from the one centre to the other enables us to compare the senses of any two pencils (see the remark at the end of §8·2). Thus we may classify congruent transformations as being *direct* or *opposite*, according to the relation between the senses of corresponding pencils. Since a reflexion reverses the sense of every pencil, a congruent transformation is direct or opposite according as it is the product of an even or odd number of reflexions. It follows from 9·63 and 9·64 that if an opposite transformation has an invariant point it must be a reflexion. Therefore the product of reflexions in any three concurrent lines is a reflexion. Thus if Ψ is a rotation about O, and Φ the reflexion in any line through O, the product $\Phi\Psi$ is another reflexion, say Φ', and

$$\Psi = \Phi\Phi\Psi = \Phi\Phi'.$$

Hence:

9·66 *Any rotation is the product of two reflexions, one of which may be the reflexion in any given line through the centre of rotation.*

In particular, the half-turn about O is the product of reflexions in any two perpendicular lines through O.

We can now prove the converse of 9·55:

9·67 *If four concurrent lines satisfy* $\langle ab \rangle \equiv \langle cd \rangle$, *then the angles* $\langle ad \rangle$ *and* $\langle bc \rangle$ *have the same bisectors.*

Proof: Let Ψ be the rotation that takes a to c, and consequently b to d, let Φ be the reflexion in one of the bisectors of $\langle bc \rangle$, and let Φ' be the reflexion $\Phi\Psi$. Then

$$\Phi = \Phi\Phi'\Phi' = \Phi\Phi\Psi\Phi\Psi = \Psi\Phi\Psi,$$

and this transforms a into

$$a^\Phi = a^{\Psi\Phi\Psi} = c^{\Phi\Psi} = b^\Psi = d.$$

Thus the given bisector of $\langle bc \rangle$ bisects $\langle ad \rangle$, too.

Interchanging b and c in 9·55, we deduce that the relation $\langle ab \rangle \equiv \langle cd \rangle$ implies $\langle ac \rangle \equiv \langle bd \rangle$. Hence:

9·68   *A rotation displaces different lines through congruent angles.*

We are now ready for an interesting specialization of Steiner's construction:

9·69   *The locus of the point of intersection of corresponding lines of two congruent pencils is a circle or a rectangular hyperbola according as the congruence is direct or opposite.**

Proof: If the two pencils are directly congruent, they are related by a rotation; thus the angle between corresponding lines is constant and the locus is a circle. On the other hand, if the two pencils are oppositely congruent, there are two pairs of corresponding lines that are parallel (given by the invariant points of the hyperbolic projectivity induced on the line at infinity). By 6·54, the locus is a conic having asymptotes in these two directions, but when corresponding lines are parallel, the respectively perpendicular lines are likewise parallel; hence the two pairs mentioned above are perpendicular, and the locus is a rectangular hyperbola.

We now possess all the material needed for Euclid's development of congruent triangles and similar triangles; e.g. two triangles are *similar* if there is a third triangle that is homothetic to the first and congruent to the second, or vice versa. Moreover, the theory of rotation leads to the *measurement* of angles, just as the theory of translation leads to the measurement of distance (§ 8·4). In fact, rotations about a point induce projectivities (resembling translations†) on the line at infinity.

EXERCISES

1. Show that our definition of a congruent transformation is redundant, since any point-to-point correspondence that preserves length also preserves collinearity.

2. Show that the product of reflexions in two parallel lines is a translation through twice the distance between the lines.

3. Deduce from 9·66 that any product of three reflexions may be regarded as the product of a reflexion and a half-turn. (It is interesting to compare this with a *rotation*, which is the product of two reflexions, and a *translation*, which is the product of two half-turns.)

4. By a second application of 9·66, show that any product of three reflexions may be regarded as a *glide reflexion*: the product of a reflexion and a translation that commute, the translation being along the axis of the reflexion.

---

* von Staudt 1847, p. 204, § 337.

† For this development, see Coxeter 1967, Chapter v. For further results on congruent transformations, see Coxeter 1991, Chapter I.

9·7   Foci. Let us now return to the theory of conics and define foci by means of a property noticed by La Hire.* A *focus* of a conic is a point at which the involution of conjugate lines coincides with the involution of perpendicular lines. For instance, the circle has a focus at its centre. (It remains to be seen whether such a point exists in any other case.) Since the involution of perpendicular lines through a point is elliptic:

9·71   *A focus is an interior point.*

(Any tangent that might be drawn through it, being self-conjugate, would have to be self-perpendicular.)

If the centre is a focus, the conic must be a circle. If not, let O be the centre and F a focus. Then OF is an axis; for the chord through F perpendicular to OF, being conjugate to OF, is bisected by OF.

9·72   *If there are two foci, their join is an axis.*

Proof: The lines through two foci F and F', perpendicular to their join FF', are both conjugate to FF'. Therefore the pole of FF' is at infinity; that is, FF' is a diameter and, consequently, an axis.

It follows that any foci which exist must all lie on one axis. In the case of a hyperbola this must, by 9·71, be the transverse axis. To establish the existence of foci (one for a parabola, one for a circle, two for a hyperbola, and two for an eccentric ellipse), we shall describe a construction for them, making use of 6·57 (the dual of Seydewitz's theorem) as applied to a triangle $TT'U$ or $a'ab$, whose vertices T and T' are joined to C (on AA') by conjugate lines p and p', as in Fig. 9·7A.

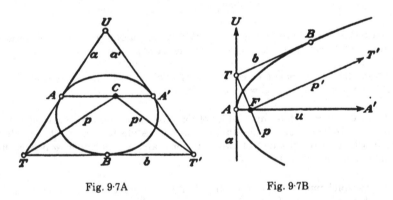

Fig. 9·7A   Fig. 9·7B

In the case of a parabola, we take a' to be the line at infinity, a the tangent at the vertex A, and b an arbitrary tangent (as in Fig. 9·7B), so

---

* For the present treatment, see von Staudt 1847, p. 208, §345 and Reye 1923, p. 155.

that any point C on the axis AA' is joined to $T = a \cdot b$ and $T' = a' \cdot b$ by conjugate lines. Then C will be a focus if these lines are perpendicular; hence we have the following construction:

9·73 *To construct the focus F of a parabola, let any tangent b meet the tangent at the vertex in T. Through T draw p perpendicular to b. Then p meets the axis in F.*

There is only one focus, for *any* focus would be joined to T by a line perpendicular to b.

In the case of a hyperbola, we take b to be an asymptote, while a and a' are the tangents at the two vertices A and A' (as in Fig. 9·7C), so that any point C on the transverse axis AA' is joined to $T = a \cdot b$ and $T' = a' \cdot b$ by conjugate lines. Then C will be a focus if these lines are perpendicular. Hence:*

9·74 *To construct the foci F and F' of a hyperbola, let either asymptote meet the tangent at either vertex in T. Then the concentric circle through T meets the transverse axis in F and F'.*

In the case of an ellipse, we take b to be the tangent at one end of the minor axis (or any other tangent not parallel to the minor axis), while a and a' are the tangents at the ends of the major axis (as in Fig. 9·7D), so that any point C on the major axis is joined to $T = a \cdot b$ and $T' = a' \cdot b$ by conjugate lines. Then C will be a focus if these lines are perpendicular. Hence:†

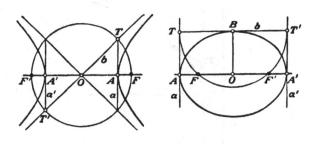

Fig. 9·7C   Fig. 9·7D

9·75 *To construct the foci F and F' of an eccentric ellipse, let either of the tangents parallel to the major axis meet the two tangents parallel to the minor axis in T and T'. Then the circle on TT' as the diameter meets the major axis in F and F'.*

To make sure that this circle will meet the major axis, we observe that its radius $BT = OA$ is greater than BO; therefore the centre O of the ellipse is

* Holgate 1930, p. 220.
† Apollonius 1891, p. 424, lib. III, Prop. 45.

interior to the circle, and AA' is a secant. If we interchanged the roles of the major and minor axes, we should have a circle of radius OB that would fail to meet a line distant OA from its centre. Thus there are no foci on the minor axis (in real geometry).

EXERCISE

Let c be a variable line through a fixed point and C its pole with respect to a given polarity. Prove that the line through C perpendicular to c envelops a parabola.

9·8　Directrices. The polar of a focus is called a *directrix*. Before considering the general conic, let us establish two interesting properties of the directrix of a parabola.

9·81　*Perpendicular tangents to a parabola meet on the directrix.*

Proof: Let a and a' be the tangents to a parabola from a point U on the directrix f, and let these meet the tangent at the vertex in T and T' (Fig. 9·8A). Since the focus F is conjugate to U, 6·57 shows that FT and FT' are conjugate; therefore they are perpendicular. FT is perpendicular to a, however, and FT' to a'. Hence $FTUT'$ is a rectangle, and a is perpendicular to a'. Since any other tangent perpendicular to a would be parallel to a' (which is impossible), this completes the proof.

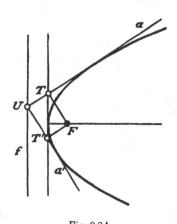

Fig. 9·8A

9·82　*The orthocentre of a triangle circumscribed to a parabola lies on the directrix.**

---

* This theorem is due to Steiner, the first proof to Holgate (1930, p. 209), and the second to J.C. Moore (see Salmon 1879, p. 247). It has seemed worth while to include both, because they are so neat.

First Proof: By the dual of 6·72, applied to the quadrangle formed by the sides of the given triangle and the line at infinity, the tangents to the parabola from the orthocentre S are a line pair belonging to the same involution as the lines joining S to the pairs of opposite vertices of this quadrilateral (which are the vertices of the triangle and the points at infinity on the opposite sides). Since these pairs of lines are perpendicular, so are the tangents.

Second Proof: Let abc be the triangle and o the line at infinity. Applying Brianchon's theorem (7·22) to the hexagon $abcc'oa'$, where c' and a' are the tangents perpendicular to c and a, we conclude that the three lines

$$(a \cdot b)(c' \cdot o), \quad (b \cdot c)(o \cdot a'), \quad (c \cdot c')(a \cdot a')$$

are concurrent. The first two are altitudes of the triangle, and the third is the directrix of the parabola.

We come now to Pappus's celebrated construction for a conic as the locus of a point whose distance from a fixed point F is e times its distance from a fixed line f, where e is a positive number called the *eccentricity*.

To establish this construction, let F be a focus, f its polar, A the vertex between F and f, B any other point on the conic, T the pole of $t = AB$, and $P = f \cdot t$ the pole of FT, as in Fig. 9·8B.* Now, FT and FP, through the focus F, are conjugate and therefore perpendicular; also, they are harmonic conjugates with respect to FA and FB (by 6·41, applied to the conjugate points $FT \cdot t$ and P on AB). Hence, by 9·53, they bisect the angle $\langle AFB \rangle$. The circle through F with centre B will meet FP again in a point C such that $BC \equiv BF$, whence, by 9·57,

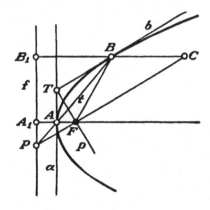

Fig. 9·8B

---

* Reye 1923, pp. 159–60, slightly simplified.

$$\langle BCF\rangle \equiv \langle CFB\rangle \equiv \langle AFC\rangle, \quad \text{that is,} \quad \langle BCP\rangle \equiv \langle AFP\rangle,$$

which shows that BC is parallel to the axis AF. Hence, if A_1 and B_1 are the points where AF and BC meet f,

$$\frac{CB}{BB_1} = \frac{FA}{AA_1}.$$

Calling this ratio e, we have $FB = CB = eBB_1$.

EXERCISES

1. Prove that the sum (or difference) of the two focal distances of a variable point on an ellipse (or hyperbola) is constant.

2. Prove that $e < 1$ for an ellipse, $e = 1$ for a parabola, $e > 1$ for a hyperbola, and $e = \sqrt{2}$ for a rectangular hyperbola.

3. Show that the conics which have a given point for focus and a given line for corresponding directrix form a self-dual system.

4. Prove that the orthocentres of the four triangles occurring in a quadrilateral are collinear.

CHAPTER 10

Continuity

The purpose of this chapter is to show how, in the presence of the axioms of incidence and order, one very simple statement about limits will suffice for the derivation of all the celebrated properties of the one-dimensional continuum, including the axioms of Archimedes and Dedekind, and Enriques's theorem (our 3·51). This treatment may be regarded as the geometrical counterpart of Weierstrass's theory of irrational numbers.

10·1 An improved axiom of continuity. Our development of real projective geometry began with five axioms of incidence (2·21–2·25 or 2·31–2·35) and six axioms of order (3.11–3.16). These were sufficient to establish many interesting theorems, such as 3·31 and 3·34; but before we could prove that a projectivity (as defined by von Staudt) is an ordered correspondence, we had to introduce a twelfth axiom, 3·51, of a decidedly complicated nature. The corresponding statements in other books are no simpler, except when algebra is used and a direct appeal is made to the system of real numbers. The following purely geometrical axiom 10·11 is so simple that it requires only eight words; however, two of the words, *monotonic* and *limit*, must be carefully defined.

A sequence of points A_0, A_1, A_2, ... is said to be *monotonic* if $A_0 A_n // A_1 A_{n+1}$ for every integer $n > 1$. (This just means that we have infinitely many points, arranged in cyclic order on the line.) The existence of such a sequence is ensured by Theorem 3·19.

A point M is called a *limit* of this sequence $\{A_n\}$ if it satisfies the following two conditions:

(1) For every integer $n > 2$, $A_1 A_n // A_2 M$.

(2) For every point P with $A_1 P // A_2 M$, there exists an n such that $A_1 A_n // PM$.

The above definitions are complete, but the following remarks (along with Fig. 10·1A) will perhaps help to clarify them. Regarding A_0 as a kind of barrier, let us say 'X precedes Y' and write $X \prec Y$, if

$$S(A_0XY) = S(A_0A_1A_2),$$

and again if $Y = A_0$. (The transitivity of this relation is a consequence of our axioms of order.) Then the sequence $\{A_n\}$ is monotonic if $A_1 \prec A_2 \prec \cdots$; the two requirements for a limit M are as follows:

(1) The points $A_1, A_2, \ldots$ all precede M.
(2) Every point P that precedes M precedes some A_n.

$$\begin{array}{cccccccc} A_0 & A_1 & A_2 & & P & A_n & M \end{array}$$

Fig. 10·1A

We are now ready for the axiom:

10·11 *Every monotonic sequence of points has a limit.*

We see at once that this limit is unique. For if M and M' are two such points, let M' precede M. By (2), M' precedes some A_n; but by (1) every A_n precedes M'. Thus the assumption $M' \prec M$ leads to a contradiction; similarly, so does the assumption $M \prec M'$.

10·2 Proving Archimedes's axiom. In order to identify the real projective line with the one-dimensional continuum described by Cantor, we must examine various properties of the continuum and see whether we can deduce them from our axioms. For instance, the property of *density*, or *internal convexity*, is a consequence of 3·19. Defining *segment* as in §3·2, we may express this property as follows:

10·21 *Every segment contains a point.*

It follows that any segment contains infinitely many points.

A more subtle property is what Hilbert[*] called the axiom of Archimedes (though it might more properly be ascribed to Eudoxus). He expressed it in affine terms, as follows: Let A_1 be any point between A and P. Take points $A_2, A_3, \ldots$ so that

$$AA_1 \equiv A_1A_2 \equiv A_2A_3 \equiv \cdots .$$

Then there exists a positive integer n such that P lies between A and A_n.

Referring to Fig. 8·4A, we see that

$$H(MA_1, AA_2), \quad H(MA_2, A_1A_3), \quad \ldots$$

whence, by 3·21, $MA_1 // AA_2$, $MA_2 // A_1A_3$, $\ldots$, i.e.

$$S(MAA_1) = S(MA_1A_2) = S(MA_2A_3) = \cdots .$$

---

[*] 1930, p. 25.

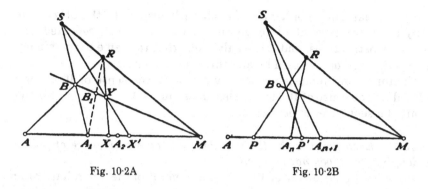

Fig. 10·2A   Fig. 10·2B

Thus the sequence $MAA_1A_2 \cdots$ is monotonic, the projectivity

$$MAA_1X \stackrel{R}{\barwedge} MBB_1Y \stackrel{S}{\barwedge} MA_1A_2X'$$

(Fig. 10·2A) is direct, and $S(MAX) = S(MA_1X')$. Using BY instead of RS, we have similarly

$$S(MAA_1) = S(MXX'),$$

which we may conveniently write as

$$X \prec X'.$$

By 10·11, the monotonic sequence $\{A_n\}$ has a limit. The axiom of Archimedes asserts that this limit is precisely M. In other words:

10·22   *If* $H(MA_n, A_{n-1}A_{n+1})$ *for all positive integers, then M is the limit of the sequence* $\{A_n\}$.

Proof: If the limit is not M, we can identify the monotonic sequence $MAA_1 \cdots$ with the $A_1A_2A_3 \cdots$ of §10·1, and the limit must be some point P' such that $A_n \prec P'$. Now construct the point $P'S \cdot MB$, which projects from R into P (so that the projectivity $X \barwedge X'$ would take P to P', as in Fig. 10·2B). We have seen that $X \prec X'$. In particular, $P \prec P'$. By the second property of a limit, there must be an n such that $P \prec A_n$. Since the projectivity is direct, it follows that $P' \prec A_{n+1}$, contradicting the first property of a limit. Hence there cannot in fact be a limit different from M.

10·3   Proving the line to be perfect.*   The points $A, A_1, A_2, \cdots$ of 10·22 form what is sometimes called a *harmonic sequence*. This is part of Möbius's *harmonic net*, or *net of rationality*, which may be described as the smallest set of points that contains, for every three of its mem-

---

* Russell 1930, p. 103; 1937, p. 291. A range is said to be *perfect* if it satisfies 10·11, 10·21 and 10·31.

bers, the harmonic conjugate of each with respect to the other two. Any three points on the line lead to a harmonic net by repeated harmonic constructions, and it is easily seen* that the same harmonic net is equally well determined by any three of its points.

Cantor's continuum is not merely dense (in the sense of 10·21) and closed (in the sense of 10·11) but also 'dense in itself': every point is the limit of a sequence. In particular:

10·31   *Each point is the limit of some monotonic sequence of points belonging to a given harmonic net.*

Proof: It is a corollary of 10·22 that any point of the given harmonic net is the limit of such a sequence, for we can construct the harmonic sequence $\{A_n\}$ from the three points M, A, A_1 that determine the harmonic net. Accordingly, let us take a point *not* belonging to the harmonic net and try to exhibit it as a limit.

Changing the notation slightly, let the given harmonic net be defined by three points A, B, C, and let Z be a point not belonging to this net. By 3·14 and 3·17, Z must occur in just one of the segments BC/A, CA/B, AB/C, say the last. For the sake of verbal economy, let us employ the language of affine geometry, regarding C as the point at infinity on the line; e.g. instead of 'the harmonic conjugate of C with respect to A and B' we say simply 'the midpoint of AB'. This midpoint belongs to the net and decomposes the segment AB (meaning AB/C) into two parts, one of which must contain Z. That part is similarly decomposed by its midpoint. We continue indefinitely in this manner, always bisecting the part that contains Z, so as to obtain a *contracting sequence of segments*, each containing the next and all containing Z. We name such a segment† LU in the order that makes $AU//LB$, so that $S(LUC) = S(ABC)$, which we write conventionally as $L \prec U$. The lower ends L and upper ends U form monotonic sequences of points, which, by 10·11, have limits M and N such that

$$L \prec M, \quad N \prec U.$$

Since $L \prec Z \prec U$, just one of the following four statements must hold:

$$M = Z \prec N, \quad M \prec Z \prec N, \quad M \prec Z = N, \quad M = Z = N.$$

We shall establish the last of these four by showing that the assumption $M \prec N$ leads to a contradiction.

Assuming $M \prec N$, let us construct points M' and N' so that

$$\mathrm{H}(CN, MM') \quad \text{and} \quad \mathrm{H}(CM, NN'),$$

---

* Veblen and Young 1910, p. 85.

† We purposely avoid the notation $L_n U_n$, because any two consecutive segments have one end in common, and we are interested in sequences of *distinct* points.

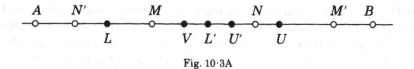

Fig. 10·3A

i.e. so that N is the midpoint of MM' and M of NN'. (This means, in effect, that the four points N', M, N, M' are 'evenly spaced', as in Fig. 10·3A.) Since M and N are the limits of lower and upper ends, we can find a lower end in the segment $N'M$ and an upper end in NM'. If these do not belong to the same one of the contracting sequence of segments, choose the latter one of the two segments involved. We thus obtain a segment LU such that

$$N' \prec L \prec M \prec N \prec U \prec M'.$$

Now construct points L', V, U' so that

$$H(CM, LL'), \quad H(CV, LU), \quad H(CN, UU'),$$

i.e. so that M, V, N are the respective midpoints of LL', LU, UU'. By 3·41 (applied to M and C, with X running from L down to N', as in Fig. 10·3B), we have $L' \prec N$; therefore $L' \prec U$. By 3·43 (applied to L and C, with X running from L' up to U, as in Fig. 10·3C), $M \prec V$. Similarly, using UU' instead of LL', we find $V \prec N$. Thus $M \prec V \prec N$. In the contracting sequence of segments, next after LU must be either LV, with $N \prec V$, or VU, with $V \prec M$. In either case we obtain a contradiction. Hence, instead of $M \prec N$, we must have $M = N$, and Z is in fact the limit of both sequences $\{L\}$ and $\{U\}$.

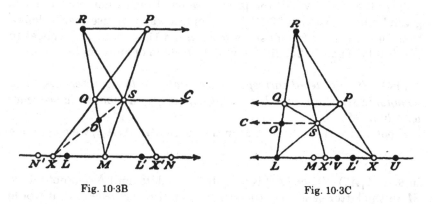

Fig. 10·3B   Fig. 10·3C

We see now that the line is dense (containing a point between any two points), closed (containing the limit of each sequence) and dense in itself (each point a limit). According to Cantor, still one more property is needed before we can be sure that this continuum is strictly similar to the class of real numbers. The final requirement is the occurrence of an enumerable 'relatively dense' subset.* A set is said to be *enumerable* if its members can be put into one-to-one correspondence with the integers (or with the rational numbers). A set of points is said to be *relatively dense* on the line if every segment contains a point of the set; in other words, there is not merely a point of the line between any two points of the special set (which follows from 10·21) but more surprisingly a point of the set between any two points of the line. This is the geometrical counterpart of the arithmetical theorem that a *rational* number can be found between any two *real* numbers.

Such an enumerable separation set is provided (for the real projective line) by a harmonic net. From the nature of its construction, this is obviously enumerable. It is relatively dense since, by 10·21 and 10·31:

10·32 *Every segment contains a point of a given harmonic net.*

This is known as the Lüroth-Zeuthen theorem† because Lüroth and Zeuthen proved it independently in 1873, using a method which resembles our proof of 10·31; however, they assumed not only 10·11 (and 3·41 and 3·43) but also Dedekind's axiom, which we shall prove in § 10·5.

It is significant that, of the four classical properties of the continuous line, the first, third, and fourth hold for the harmonic net itself. Thus the second ('closure') is the crucial property that distinguishes the continuum, and naturally this is the one that has to be taken as an axiom.

10·4 The fundamental theorem of projective geometry. The essential steps in our proof of the fundamental theorem 4·21 were 3·51, 3·62, 4·11, 4·12. How will the procedure be altered when continuity is given by 10·11 instead of 3·51? The simplest way is to use Pieri's definition for a segment (p. 38) so as to obtain 4·11 without any appeal to continuity. Then 4·12 will follow with the help of one simple lemma:

10·41 *If an ordered correspondence relates A_n to A'_n, where $\{A_n\}$ is a monotonic sequence with limit M, then $\{A'_n\}$ is a monotonic sequence with limit M'.*

Proof: Since $A_1 \prec A_2 \cdots \prec M$ in the sense $S(A_0 A_1 A_2)$, we must have

$$A'_1 \prec A'_2 \cdots \prec M'$$

in $S(A'_0 A'_1 A'_2)$. Hence $\{A'_n\}$ is monotonic, and its limit N' cannot follow M' in the latter sense. On the other hand, if N' preceded M', it would come from a point N separated from M by some A_n, implying

---

* Forder 1927, p. 14. This is called a *median class* in Russell 1937, p. 104.
† See Whitehead 1906, pp. 30–3, or Mathews 1914, pp. 43–6.

$$N' \prec A'_n \prec M',$$

whereas every A'_n should precede the limit N'. Hence in fact N' must coincide with M'.

To prove 4·12, we recall that a projectivity is a correspondence that preserves the harmonic relation. Thus, if three points are invariant, the whole harmonic net determined by those three points must be invariant. Hence, by 4·11, 10·41, and 10·31, *every* point is invariant.

10·5 Proving Dedekind's axiom. Dedekind's axiom may be expressed as follows:

10·51 *For every division of all the points of a given segment or interval α into two non-empty sets R_1 and R_2, such that every point of R_1 precedes every point of R_2, there exists a point M in α which has the property that every point of α preceding M belongs to R_1 and every point of α following M belongs to R_2.*

In other words, if $L \prec U$ for every point L of the lower set and every point U of the upper, then one of the two sets contains a dividing point M such that $L \prec M \prec U$, except that either L or U might coincide with M.

The proof closely resembles that of 10·31. Let α be AB/C or $\overline{AB}/C$, and consider its midpoint, i.e. the harmonic conjugate of C with respect to A and B. Call the midpoint L or U according as it belongs to R_1 or R_2. Similarly bisect LB or AU, as the case may be. In this manner we obtain a contracting sequence of segments, each having one end in R_1 and the other in R_2. We see, as before, that the lower ends and upper ends have the same limit M, which is Dedekind's dividing point.

10·6 Enriques's theorem. We are now ready to prove the theorem that we regarded as an axiom in 3·51. We take first the case when the correspondence is direct:

10·61 *If a direct correspondence relates an interval $\overline{AB}/C$ to an interior interval $\overline{A'B'}/C$, then the latter contains an invariant point M such that there is no invariant between A and M (in $\overline{AB}/C$).*

Proof: If A is invariant, there is no more to be said—M coincides with A. If not, suppose A is related to another point A', A' to A'', and so on. Then the iterated correspondence provides a monotonic sequence $AA'A'' \cdots$, whose limit M is invariant by 10·41 (applied to the sequences $AA'A'' \cdots$ and $A'A''A''' \cdots$). A point between A and M cannot be invariant, for if it coincides with some $A^{(n)}$, it is related to the different point $A^{(n+1)}$, and if it lies in a segment $A^{(n-1)}A^{(n)}$, its corresponding point lies in the different segment $A^{(n)}A^{(n+1)}$.

This is simple. The slightly more difficult case is when the correspondence is opposite:

10·62 *If an opposite correspondence relates an interval $\overline{AB}/C$ to an interior interval $\overline{A'B'}/C$, then the latter contains an invariant point.*

Proof:[*] Assuming that there is no invariant point, we derive a contradiction by the following argument: We use Dedekind's axiom 10·51, taking the lower set to consist of those points of $\overline{AB}/C$ which *precede* their corresponding points, while the upper set consists of those which *follow* their corresponding points. These sets are easily seen to have the requisite properties. First, the sets are not empty; for since $B' \prec A'$, the lower contains A and the upper B. Second, if L precedes its corresponding point L' while U follows U', L must precede U; for otherwise we should have $U' \prec U \prec L \prec L'$, so that LU and $L'U'$ would have the same sense. Dedekind's axiom yields a dividing point M, such that every point preceding M is an L and every point following M is a U.

Applying the given correspondence to the relations

$$L \prec L', \quad U' \prec U,$$

we obtain

$$L'' \prec L', \quad U' \prec U'';$$

thus every L' is a U and every U' is an L. Applying the correspondence also to M (which, by our assumption, is not invariant), we find that, in the interior interval $\overline{A'B'}/C$, every point following M' is an L' and every point preceding M' is a U'; thus every point preceding M' is an L and every point following M' is a U.

Hence $M' = M$, contradicting our assumption that no point is invariant. Therefore some point must be invariant (and it follows, as in §3·5, that the invariant point is unique).

---

[*] The proof given in the first edition of this book, along the lines of Enriques 1930, pp. 71–5, has been notably simplified by J.R. Vanstone.

CHAPTER 11

The Introduction of Coordinates

In Chapter 10 we discussed many properties of the real projective line, but there remain certain questions that would be difficult, if not impossible, to answer without using the concept of a coordinate or abscissa. For instance, how can you be sure that a harmonic net does not exhaust all the points on the line?

We saw, in §8·4, that pairs of points on the affine line belong to an involution if their algebraic distances from a fixed point on the line have either a constant sum or a constant product. The question now arises: What is the projective counterpart of this affine statement? More precisely: What projective entities can be added or multiplied? One answer was given by von Staudt,* who used sets of four points, which he called *Würfe* (i.e. 'throws' or 'casts'). Hessenberg, in 1905, simplified that treatment by fixing three of the four points and operating with the remaining one. Instead of adding or multiplying segments OX, as in the affine line, we now add or multiply points X in the presence of three fixed points, which play the role of the numbers 0, 1, ∞. (Anyone familiar with the vectorial approach to analytic geometry will understand how a vector OX and its end point X are for many purposes interchangeable.)

Following O'Hara and Ward,† we develop this theory 'in one dimension', using elementary properties of involutions (instead of constructions involving arbitrary points outside the given line). Although this method obscures Hilbert's famous discovery of the connexion between Pappus's theorem and the commutativity of multiplication, it has the advantage of allowing the range of points to be on a conic just as well as on a line. We shall see, in Figs. 11·1A and 11·2A, how every easily the sum or product of two points on a conic may be constructed.

The chief novelty arises in §11·8, where we introduce two-dimen-

---

* 1857, pp. 166–94.
† 1937, pp. 155–6. Cf. Veblen and Young 1910, pp. 141–56, 232.

sional homogeneous coordinates. The use of a conic makes it unnecessary to mention either cross ratio or non-homogeneous coordinates.

11·1 Addition of points. Relative to two fixed points P_0 and P_∞ on a given line or conic, we define the *sum $A + B$* of two arbitrary points (on the same line or conic, but distinct from P_∞) to be the companion of P_0 in the hyperbolic involution

$$(AB)(P_\infty P_\infty).$$

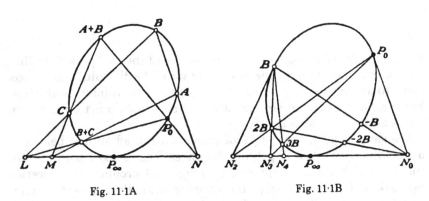

Fig. 11·1A   Fig. 11·1B

Thus $A + P_0 = A$, and $A + A$ is the harmonic conjugate of P_0 with respect to A and P_∞. The commutative law

$$A + B = B + A$$

is satisfied immediately, and the equation $X + B = A$ can be solved by taking X to be the companion of B in $(AP_0)(P_\infty P_\infty)$. In particular, $-B$ (such that $-B + B = P_0$) is the harmonic conjugate of B with respect to P_0 and P_∞.

To establish the associative law

$$A + (B + C) = (A + B) + C,$$

we observe that P_∞ is an invariant point of both the involutions

$$(A\,B)(A + B\,P_0) \quad \text{and} \quad (C\,B)(B + C\,P_0),$$

whence, by 4·73, P_∞ is also an invariant point of

$$(A\,B + C)(A + B\,C).$$

Thus A and $B + C$ have the same sum as $A + B$ and C.*
We now define $2B = B + B$, $3B = 2B + B$, and so on. Since

$$(m - 1)B + (m + 1)B = mB + mB,$$

----

* This method is due to Alex Rosenberg.

each successive multiple $(m + 1)B$ arises as the harmonic conjugate of $(m - 1)B$ with respect to mB and P_∞.

The operation of adding B can be expressed as a projectivity. In fact, the above definitions imply

$$P_\infty P_0(-B)A \barwedge P_\infty P_0 B(-A) \barwedge P_\infty BP_0(A + B).$$

If B is a fixed point (not P_0 or P_∞), this combined projectivity is independent of the choice of A, and hence it relates a variable point X to $X + B$. This still holds when X is P_∞, provided that we extend the definition of addition* by declaring that, if $B \neq P_\infty$, $P_\infty + B = P_\infty$.

By 3·22 and 3·34, the projectivity

11·11 $P_\infty P_0(-B) \barwedge P_\infty BP_0$

is direct. (In fact, since P_∞ is the only invariant point, it is parabolic.) Applying it repeatedly to any point (other than P_∞), we obtain a monotonic sequence. In particular:

11·12   *The sequence of points*

$$\ldots, -3B, -2B, -B, P_0, B, 2B, 3B, \ldots$$

is monotonic.

In other words, the relation $S(kB\ mB\ P_\infty) = S(P_0\ B\ P_\infty)$ holds if and only if $k < m$.

EXERCISES

1. Derive 11·12 from the fact that mB and P_∞ separate $(m \pm 1)B$.

2. Show that the sum of two points A and B on a conic may be constructed as in Fig. 11·1A, AB meets the tangent at P_∞ in N (the centre of the additive involution), and $A + B$ is the point where the line NP_0 meets the conic again. Deduce that $P_\infty + B = P_\infty$ $(B \neq P_\infty)$.

3. Derive the associative law for addition from Pascal's theorem (our 7·21) applied to the hexagon $ABC(A + B)(P_0(B + C)$.

4. Let $P_0 B$ (Fig. 11·1A or B) meet the tangent at P_∞ in N_1. Show that the point of contact of the second tangent from N_1 is a point $\frac{1}{2}B$ such that $\frac{1}{2}B + \frac{1}{2}B = B$.

5. Given the parabola in its familiar Cartesian form $y = x^2$, proves that the point on it with abscissa $a + b$ may be located by drawing through the vertex $(0, 0)$ the

---

* This convention can be justified by appealing to the degenerate involution

$$(P_\infty B)(P_\infty P_\infty),$$

which relates every point (in particular, P_0) to P_∞.

chord parallel to $(a, a^2)(b, b^2)$, as in Fig. 11·1C. Show how this agrees with the formal addition of points on the parabola.

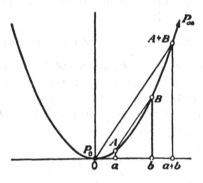

Fig. 11·1C

11·2   Multiplication of points. Relative to three fixed points P_0, P_1, P_∞, on a given line or conic, we define the *product* $A \times B$ of two arbitrary points (on the same line or conic, but distinct from P_0 and P_∞) to be the companion of P_1 in the elliptic or hyperbolic involution

$$(AB)(P_0 P_\infty).$$

For example, $A \times P_1 = A$. The commutative law, $A \times B = B \times A$, is satisfied immediately, and the equation $X \times B = A$ can be solved by taking X to be the companion of B in $(AP_1)(P_0 P_\infty)$. In particular, B^{-1} (such that $B^{-1} \times B = P_1$) is the companion of B in the hyperbolic involution $(P_1 P_1)(P_0 P_\infty)$, whose second invariant point is $-P_1$, or say P_{-1}; thus B^{-1} is the harmonic conjugate of B with respect to P_1 and P_{-1}.

To establish the associative law

11·21 $$A \times (B \times C) = (A \times B) \times C,$$

we observe that $P_0 P_\infty$ is a pair of both the involutions

$$(A B)(A \times B P_1) \quad \text{and} \quad (C B)(B \times C P_1),$$

whence, by 4·68, $P_0 P_\infty$ is also a pair of $(A B \times C)(A \times B C)$. Thus A and $B \times C$ have the same product as $A \times B$ and C.

Since

$$A^{-1} \times B^{-1} \times B \times A = P_1,$$

we have

$$A^{-1} \times B^{-1} = (B \times A)^{-1} = (A \times B)^{-1}.$$

The operation of multiplying by B can be expressed as a projectivity. By 2·71 and the definition of $A \times B$,

$$P_0 P_\infty P_1 A \;\overline{\wedge}\; P_\infty P_0 A P_1 \;\overline{\wedge}\; P_0 P_\infty B(A \times B).$$

If B is a fixed point (not P_0 or P_∞), this combined projectivity is independent of the choice of A and hence it relates a variable point X to $X \times B$. This still holds when X is P_0 or P_∞, provided that we extend the definition of multiplication* by declaring that, if $B \neq P_\infty$, $P_0 \times B = P_0$ and, if $B \neq P_0$, $P_\infty \times B = P_\infty$.

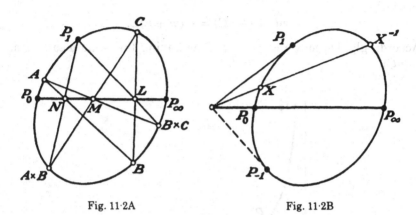

Fig. 11·2A   Fig. 11·2B

This suggests an alternative proof for the associate law 11·21. Using the projectivity

11·22 $P_0 P_\infty P_1 \,\overline{\wedge}\, P_0 P_\infty C$

which relates every X to $X \times C$, we have

$$P_0 P_\infty (A \times B)A \,\overline{\wedge}\, P_\infty P_0 P_1 B \,\overline{\wedge}\, P_\infty P_0 C(B \times C).$$

Thus A and $B \times C$ have the same product as $A \times B$ and C.

Similarly, to prove the distributive law

11·23 $(A \times C) + (B \times C) = (A + B) \times C,$

we observe that, from the definition of $A + B$,

$$P_\infty ABP_0 \,\overline{\wedge}\, P_\infty BA(A + B),$$

whence, by 11·22,

$$P_\infty (A \times C)(B \times C)P_0 \,\overline{\wedge}\, P_\infty (B \times C)(A \times C)[(A + B) \times C].$$

This involution exhibits $(A + B) \times C$ as the sum of $A \times C$ and $B \times C$.

By repeated application of 11·23, we see that, for any positive integer n, $n(A \times C) = (nA) \times C$. In particular,

$$nC = P_n \times C, \quad \text{where} \quad P_n = nP_1.$$

---

* These conventions can be justified by means of the degenerate involutions

$$(P_0 B)(P_0 P_\infty) \quad \text{and} \quad (P_\infty B)(P_\infty P_0).$$

If m is the greater of two positive integers m and n, we have

$$-mC + nC + (m - n)C = -mC + mC = P_0,$$

whence

$$-mC + nC = -(m - n)C;$$

similarly

$$-mC + (-nC) = -(m + n)C.$$

Accordingly, we may regard $(-n)C$ as having the same meaning as $-(nC)$.

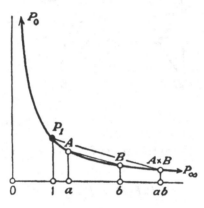

Fig. 11·2C

EXERCISES

1. Verify that $(-A) \times (-B) = A \times B$. (*Hint:* Use §4·6, Ex. 3.)

2. Show that the product of two points A and B on a conic may be constructed as in Fig. 11·2A; AB meets $P_0 P_\infty$ in N (the centre of the multiplicative involution), and $A \times B$ is the point where the line NP_1 meets the conic again. Deduce that

$$P_0 \times B = P_0 \quad (B \neq P_\infty) \quad \text{and} \quad P_\infty \times B = P_\infty \quad (B \neq P_0).$$

3. Derive the associative law for multiplication from Pascal's theorem (our 7·21), applied to the hexagon $ABC(A \times B)P_1(B \times C)$.

4. Verify that the projectivity 11·22 is direct if P_0 and P_∞ do not separate P_1 and C (e.g. if these points are arranged as in Fig. 11·2A). In this case $P_0 P_\infty$ meets $P_1 C$ in an exterior point, from which we can draw two tangents to the conic. Show that their points of contact are $\pm C^{1/2}$, such that $C^{1/2} \times C^{1/2} = C$.

5. Given the rectangular hyperbola $xy = 1$ or $y = x^{-1}$, prove that the point with abscissa ab may be located by drawing through the vertex $(1, 1)$ the chord parallel to $(a, a^{-1})(b, b^{-1})$, as in Fig. 11·2C. Show how this agrees with the multiplication of points on the rectangular hyperbola.

11·3 Rational points. Defining

$$P_n = nP_1, \quad P_{-n} = -P_n, \quad P_{1/n} = (P_n)^{-1},$$

$$P_{m/n} = mP_{1/n}, \quad P_{-m/n} = -P_{m/n},$$

we obtain a definite point P_a for every rational number a. We call P_a a *rational point* and a its *abscissa*.

The addition and multiplication of such points agree with the addition and multiplication of the corresponding numbers. For if $a = k/n$ and $b = m/n$ (where k and m are integers, while the common denominator n is a positive integer), we have

$$P_a + P_b = kP_{1/n} + mP_{1/n} = (k + m)P_{1/n} = P_{(k+m)/n} = P_{a+b}.$$

Again, if $a = k/l$ and $b = m/n$ (where l and n are positive integers),

$$P_a \times P_b = P_k \times P_{1/l} \times P_m \times P_{1/n} = P_{km} \times P_{1/l} \times P_{1/n}$$

$$= kmP_{1/ln} = P_{km/ln} = P_{ab}.$$

Moreover, the *order* of the points P_a agrees with the order of the rational numbers a; for to see whether P_a precedes or follows P_b, we express a and b in terms of a common denominator, say $a = k/n$, $b = m/n$, and observe where P_a and P_b occur in the sequence 11·12 with $B = P_{1/n}$. We conclude that

$$S(P_a P_b P_\infty) = S(P_0 P_1 P_\infty)$$

if and only if $a < b$.

11·4 Projectivities. Setting $B = P_b$ in 11·11, we obtain a projectivity that relates a variable point P_x to P_{x+b}. Thus the transformation of abscissae

$$x' = x + b$$

represents a projectivity $P_x \barwedge P_{x+b}$, which is parabolic if $b \neq 0$.

Similarly, setting $C = P_a$ in 11·22, we obtain a projectivity that relates P_x to P_{ax}. Thus the transformation

$$x' = ax \quad (a \neq 0)$$

represents a projectivity $P_x \barwedge P_{ax}$, which is hyperbolic if $a \neq 1$.

The product of these two elementary transformations is

11·41 $$x' = ax + b \quad (a \neq 0).$$

The third kind of elementary transformation

$$x' = \frac{1}{x}$$

represents the hyperbolic involution $(P_0 P_\infty)(P_1 P_1)$, as in Fig. 11·2B.

By judiciously combining all three elementary transformations, we obtain the linear fractional transformation

$$x' = \frac{a}{c} + \frac{b - ad/c}{cx + d} = \frac{ax + b}{cx + d},$$

where $c \neq 0$ and $b \neq ad/c$. These inequalities can we weakened to

$$ad \neq bc;$$

for by setting $c = 0$ (and $d = 1$) we obtain 11·41. Since $P_\infty^{-1} = P_0$, we can take care of the possibility that $x = \infty$ by writing the linear fractional transformation in the two alternative forms

11·42 $x' = \dfrac{ax + b}{cx + d} = \dfrac{a + b/x}{c + d/x}$ $(ad - bc \neq 0)$.

Another way of writing it is

$$cxx' - ax + dx' - b = 0.$$

The projectivity thus represented is an involution if x and x' are interchangeable, i.e. if $-a = d$. In particular, the involution with invariant points P_a and P_b is represented by

$$xx' - \tfrac{1}{2}(a + b)(x + x') + ab = 0.$$

Hence:

11·43   *The relation* $\mathrm{H}(P_a P_b, P_c P_d)$ *is equivalent to*

$$(a + b)(c + d) = 2(ab + cd).$$

EXERCISES

1. Show that the parabolic projectivity 11·11, as applied to points on a conic, has the tangent at P_∞ for its axis.

2. Show that the hyperbolic projectivity 11·22 (with $C \neq P_1$) has the secant $P_0 P_\infty$ for its axis.

3. Show that the projectivity 11·41 is hyperbolic except when $a = 1$ and that it is direct or opposite according as a is positive or negative.

4. Show that the transformation $x' = ax$ can be obtained by judiciously combining $x' = b + x$ and $x' = 1/x$.

Hint: $-n^2 x = n + \dfrac{1}{m+} \dfrac{1}{n+} \dfrac{1}{x}$,   where $m = -\dfrac{1}{n}$.

---

* Mendelsohn 1944.

11·5 The one-dimensional continuum. The various steps by which we have derived the general rational point P_a from three arbitrary points P_0, P_1, P_∞ may all be expressed in terms of harmonic conjugacy:

$$H(P_1 P_\infty, P_0 P_2), \qquad H(P_m P_\infty, P_{m-1} P_{m+1}), \qquad H(P_0 P_\infty, P_m P_{-m}),$$

$$H(P_1 P_{-1}, P_n P_{1/n}), \quad H(P_{m/n} P_\infty, P_{(m-1)/n} P_{(m+1)/n}), \quad H(P_0 P_\infty, P_a P_{-a}).$$

Conversely, by 11·43, the harmonic conjugate of P_c with respect to P_a and P_b is another rational point P_d. Hence:

11·51 *The rational points P_a, along with P_∞, form a harmonic net.*

(For this reason, a harmonic net is sometimes called a *net of rationality*.)

We are now ready to show how the remaining points of the range may be included in this algebraic treatment by defining irrational abscissae.

Let any real number x be expressed as the limit of a monotonic sequence of rational numbers a. By Axiom 10·11 (which naturally holds on the conic just as well as on the line), the corresponding sequence of points P_a has a definite limit, which we denote by P_x.

Conversely, by 10·31, any given point on the line or conic may be regarded as the limit of a monotonic sequence of rational points. The corresponding sequence of rational numbers a is eventually monotonic in the algebraic sense (after possibly discarding some initial terms of the wrong sign); thus it is either divergent or convergent. In the former case, the given point must have been P_∞; in the latter, the limit of the a's is a real number x, and the point is P_x.

The number x, whether real or infinite, is called the *abscissa* of P_x. Thus, when the three fundamental points P_0, P_1, P_∞ have been assigned, every point on the line or conic has a uniquely determined abscissa.

Since the number of real numbers is strictly greater than the number of rational numbers,[*] we see at last why it is that the harmonic net constitutes only an 'infinitesimal part' of the whole range.

We can now remove the restriction to rational abscissae in § 11·4. With real numbers a, b, c, d, 11·42 is the most general projectivity; for[†] $P_0 P_1 P_\infty \,\overline{\wedge}\, P_p P_q P_r$ is given by

$$x' = \frac{r(p - q)x + p(q - r)}{(p - q)x + (q - r)},$$

which is a valid transformation provided that p, q, r are all different.

---

[*] Russell 1930, pp. 85–6.
[†] Veblen and Young 1910, p. 155.

Hence, if a variable point P_x (on a line or a conic) is projectively related to $P_{x'}$ (on the same line or conic or another), then the abscissae x and x' must be connected by a linear fractional transformation. In particular, the most general projectivity preserving P_∞ is 11·41 (which takes P_0 to P_b and P_1 to P_{a+b}).

Sine $P_0 P_1 P_\infty P_x \barwedge P_0 P_a P_\infty P_{ax}$, any construction by which P_x is derived from $P_0 P_1 P_\infty$ will yield P_{ax} when applied to $P_0 P_a P_\infty$. Thus P_x can be renamed P_{ax} provided P_1 is renamed P_a. (The new P_1 is the old $P_{1/a}$.) More generally, instead of regarding the transformation 11·42 as a projectivity, we may equally well regard it as a consistent renaming of all the points (without altering any of their geometrical properties). Such renaming is called a *change of scale*.

We have seen (§ 11·4) that the general involution is

$$cxx' - a(x + x') - b = 0 \quad (a^2 + bc \neq 0).$$

This reduces to $x + x' = k$ when $c = 0$ and to

$$(x - a)(x' - a) + g = 0 \quad (g \neq 0)$$

otherwise. By a simple change of scale (namely, $x \to x + \tfrac{1}{2}k$ or $x + a$) these relations become

11·52 $x + x' = 0$

and $xx' + g = 0$. The former, having invariant points P_0 and P_∞, may be taken as the canonical form for a hyperbolic involution. The latter is elliptic if $g > 0$, in which case we can make further simplification by the change of scale $x \to x\sqrt{g}$. Thus the canonical form for an elliptic involution is

11·53 $xx' + 1 = 0.$

EXERCISES

1. The harmonic net based on three collinear points A, B, C naturally contains D, the harmonic conjugate of C with respect to A and B. Show that it does *not* contain the invariant points of the hyperbolic involution $(AD)(BC)$. (*Hint:* Take the abscissae of A, B, C, D to be 0, 2, 1, ∞.)

2. If to the harmonic net we adjoin such further points arising from every set of four points already obtained, have we then exhausted all the points on the line?

3. Show how § 7·5, Ex. 5, would enable us to express any elliptic involution in the form $(P_0 P_\infty)(P_1 P_{-1})$, thereby justifying 11·53 immediately.

4. With any two points on a line (or conic) we may associate the quadratic equation whose roots are their abscissae. If three point pairs form a quadrangular set (§ 4·7), prove that their equations

$$p_i x^2 + q_i x + r_i = 0 \quad (i = 1, 2, 3)$$

satisfy

$$\begin{vmatrix} p_1 & q_1 & r_1 \\ p_2 & q_2 & r_2 \\ p_3 & q_3 & r_3 \end{vmatrix} = 0.$$

(*Hint*: If the point pairs belong to the involution

$$b + a(x + x') - cxx' = 0,$$

we have

$$bp_i - aq_i - cr_i = 0.)$$

11·6 Homogeneous coordinates. Given any five points of which no three are collinear, we can draw a definite conic through them, take an arbitrary sixth point P_1 on the conic, and name two of the given points P_0 and P_∞. Then the remaining three points have definite abscissae x_1, x_2, x_3; and a different choice of the sixth point P_1 would have the effect of multiplying all of x_1, x_2, x_3 by some number a.

We call x_1, x_2, x_3 the *coordinates* of the point P_∞ for the *triangle of reference* $P_{x_1}P_{x_2}P_{x_3}$ and *unit point* P_0. From the above remarks we see that these coordinates are quite definite, apart from the possibility of multiplying all three by the same constant; in other words, they are *homogeneous* coordinates. We denote P_∞ by the symbol (x_1, x_2, x_3) with the understanding that for any non-zero a, the symbol (ax_1, ax_2, ax_3) denotes the same point.

The above definition implies that the numbers x_1, x_2, x_3 are all distinct and different from zero. We shall see later that this restriction can be removed, except that they must not be all zero. In particular, the name *unit point* will be justified when we have found the coordinates 1, 1, 1 for P_0.

It should be noticed that to find the coordinates for a new point, we generally have to start over again with a new conic. Moreover, we cannot yet construct a point with given coordinates.

11·7 Proof that a line has a linear equation. By projecting the points of a conic from a fixed point (on the conic) onto a line, we obtain points (on the line) which may be regarded as having the same abscissae, referred to the projections of the fundamental points P_0, P_1, P_∞. In particular, by projecting the points P_0, P_∞, P_{x_1}, P_{x_2}, P_{x_3} of § 11·6 from P_{x_2} on the line $P_{x_1}P_{x_3}$ and from P_{x_1} onto $P_{x_2}P_{x_3}$, we obtain points whose abscissae on those sides of the triangle (left and right) are as indicated in Fig. 11·7A. Applying the changes of scale

$$x \rightarrow \left(\frac{x}{x_1} - 1\right) \Big/ \left(\frac{x}{x_3} - 1\right) = \left(\frac{1}{x_1} - \frac{1}{x}\right) \Big/ \left(\frac{1}{x_3} - \frac{1}{x}\right)$$

and

$$x \rightarrow \left(\frac{x}{x_2} - 1\right) \Big/ \left(\frac{x}{x_3} - 1\right) = \left(\frac{1}{x_2} - \frac{1}{x}\right) \Big/ \left(\frac{1}{x_3} - \frac{1}{x}\right)$$

to the respective sides, we obtain the revised abscissae indicated in Fig. 11·7B.

Let any line through P_∞ meet these sides in the points u_1 and u_2. Since

$$\infty \; 0 \; u_1 \; \frac{x_3}{x_1} \overset{P_\infty}{\underset{\wedge}{=}} \infty \; \frac{x_3}{x_2} \; u_2 0,$$

the connexion between u_1 and u_2 is of the form 11·41, viz., since the values 0 and x_3/x_1 for u_1 correspond to the values x_3/x_2 and 0 for u_2,

11·71 $$x_1 u_1 + x_2 u_2 - x_3 = 0.$$

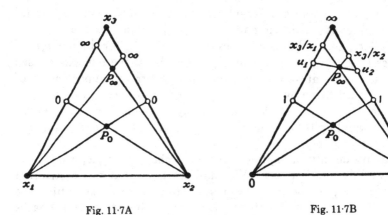

Fig. 11·7A   Fig. 11·7B

We can fix this line by fixing the values of the two numbers u_1 and u_2, which determine definite points on those two sides of the triangle in terms of 0, 1, ∞. Now let the point P_∞ vary on the line. This means that the numbers x_1, x_2, x_3 will vary; but they will continue to satisfy 11·71, which may thus be regarded as the *equation* of the line. This merely means that it is condition for the variable point (x_1, x_2, x_3) to lie on the line.

The point of intersection of two such lines is obtained by solving the two simultaneous equations for $x_1 : x_2 : x_3$. In this way we may extend the definition of coordinates to points on the sides of the triangle of reference or on the lines joining the vertices to the unit point P_0. Thus the point marked u_2 in Fig. 11·7B is where the line 11·71 meets another such line with a different u_1, namely,

$$(0, 1, u_2);$$

e.g. the point marked 1 on that same side is $(0, 1, 1)$. We now see that the characteristic property of a point on the side $P_{x_2}P_{x_3}$ of the triangle of reference is the vanishing of the first coordinate; hence the three sides have the equations

$$x_1 = 0, \quad x_2 = 0, \quad x_3 = 0.$$

Consequently the vertices, where these sides meet in pairs, are

$$(1, 0, 0), \quad (0, 1, 0), \quad (0, 0, 1).$$

Setting $u_1 = 0$ in 11·71, we obtain the line $u_2 x_2 - x_3 = 0$, which joins $(1, 0, 0)$ to $(0, 1, u_2)$. This holds for all values of u_2 except, possibly, $u_2 = 1$; accordingly, we extend the meaning of coordinates so as to make it hold there, too. Thus the lines joining P_0 to the vertices of the triangle have the equations

$$x_2 = x_3, \quad x_3 = x_1, \quad x_1 = x_2,$$

and the point P_0 itself is $(1, 1, 1)$.

We now see how to obtain coordinates for any given point. Conversely, given three real numbers x_1, x_2, x_3, not all zero, we can locate the point (x_1, x_2, x_3) as follows: If two of the coordinates are zero, the point is a vertex. If one is zero, the point lies on a side; for example, $(0, x_2, x_3)$ has abscissa x_3/x_2 referred to

$$P_0 = (0, 1, 0), \quad P_1 = (0, 1, 1), \quad P_\infty = (0, 0, 1).$$

If none is zero, we join $(1, 0, 0)$ to $(0, x_2, x_3)$ and $(0, 1, 0)$ to $(x_1, 0, x_3)$, locating (x_1, x_2, x_3) as the point where these joins intersect.

11·8 Line coordinates. Finally, we restore the symmetry of the three coordinates by writing 11·71 in the homogeneous form

$$x_1 X_1 + x_2 X_2 + x_3 X_3 = 0,$$

or

11·81
$$X_1 x_1 + X_2 x_2 + X_3 x_3 = 0,$$

and we call the coefficients X_i the *coordinates* (line coordinates, envelope coordinates, or tangential coordinates) of the line

$$[X_1, X_2, X_3].$$

This device enables us to interchange points and lines in accordance with the principle of duality. The equation 11·81 is essentially self-dual, being the condition for the line $[X_1, X_2, X_3]$ and point (x_1, x_2, x_3) to be incident. If we fix the point instead of the line, it is the condition for a variable line to pass through a fixed point; i.e. the point (x_1, x_2, x_3) has the equation

$$x_1 X_1 + x_2 X_2 + x_3 X_3 = 0,$$

and its coordinates are the coefficients of X_1, X_2, X_3 in its equation. In particular, the vertices of the triangle of reference and the unit point have the equations

$$X_1 = 0, \quad X_2 = 0, \quad X_3 = 0, \quad X_1 + X_2 + X_3 = 0.$$

In terms of line coordinates, the sides are

$$[1, 0, 0], \quad [0, 1, 0], \quad [0, 0, 1],$$

and the lines joining the vertices to the unit point are

$$[0, 1, -1], \quad [-1, 0, 1], \quad [1, -1, 0].$$

EXERCISES

1. Show that the unit line $[1, 1, 1]$ meets the sides of the triangle of reference in the points $(0, 1, -1)$, $(-1, 0, 1)$, $(1, -1, 0)$.

2. Show that the two points $(0, 1, \pm x)$ are harmonic conjugates with respect to $(0, 1, 0)$ and $(0, 0, 1)$. Deduce the condition $x_1 X_1 = x_2 X_2 = x_3 X_3$ for the point (x_1, x_2, x_3) and line $[X_1, X_2, X_3]$ to be trilinear pole and polar with respect to the triangle of reference.

3. Verify the result of Ex. 2 without using harmonic conjugates, by obtaining the coordinates of the various points and lines in Fig. 7·7A, beginning with

$$A = (1, 0, 0), \quad B = (0, 1, 0), \quad C = (0, 0, 1), \quad X = (x_1, x_2, x_3).$$

CHAPTER 12

The Use of Coordinates

In Chapter 11 we saw how a system of coordinates is inherent in synthetic geometry. In the present chapter we shall reverse the process, building up the analytic geometry from first principles, and deriving the theorems (including the axioms) from properties of numbers. We shall find that the analytic method enables us to solve some problems more easily. On the other hand, it would be a grave mistake to abandon the synthetic method, which is far more stimulating to one's geometrical ingenuity.

12·1 Consistency and categoricalness.* In Chapters 2–7 and 10 we developed the geometry of the real projective plane as a logical system based on the primitive concepts *point, line, incidence, separation*, and the twelve axioms 2·21–2·25, 3·11–3·16, and 10·11. This system has two essential properties: it is consistent and it is categorical. Before attempting to define these terms, let us remark that the properties are tested by means of *models*, wherein the primitive concepts, instead of remaining undefined, are defined in terms of concepts sufficiently familiar to be taken for granted. To establish the validity of a model, we merely have to verify that the given definitions for the primitive concepts enable us to *prove* the axioms.

When we say that a logical system is *consistent*, we mean that it cannot lead (by any chains of deduction, however long) to two contradictory propositions. The existence of a single model suffices to establish consistency; for any two contradictory results would imply contradictory properties of the model, and the absurdity would be manifest. The chief difficulty is to find an entirely satisfactory model.

Many would be prepared to take for granted, as a matter of experience, the ordinary geometry of Euclid or the affine geometry that can

---

* Cf. Veblen and Young 1910, pp. 1–6. The relation between synthetic and analytic geometry has been very ably described by Robson (1940, Chapter 8; 1947, Chapter 19).

be extracted from it. Then we can define ideal elements as in § 1·4 and verify that the affine plane, plus its points at infinity and line at infinity, forms a model for the projective plane.

Others might reject this model on the grounds that the space of experience is only approximately Euclidean. Setting aside the question as to whether a straight line is better approximated by a taut string or a ray of light, they would argue that Euclid's postulate of parallelism (which is an essential part of affine geometry) can be tested experimentally only in a neighbourhood that is very small from the astronomical standpoint. Such persons might be prepared to take for granted the *local* properties of ordinary space. Then a model for the real projective plane is provided by the lines and planes through a fixed point in space. These lines and planes represent the points and lines of the projective plane, while incidence and order retain their customary meaning. This model has the great advantage of symmetry: there is no 'line at infinity' to play a special role.

Still others might object that even this symmetrical model rests on intuitive ideas of space that cannot be justified by purely logical means. For them we must devise a model in which every geometrical concept is defined in terms of numbers. The validity of such an analytic model will be verified in §§ 12·3 and 12·4. Of course, there remains the question of the consistency of the number system, but at that stage the geometer delegates his responsibility to the algebraist.

When we say that a logical system is *categorical*, we mean that it is unique, in the sense that any model is isomorphic with any other. Thus a geometry is categorical if the entities which represent all the points and lines in one model can be put into correspondence (one-to-one) with those which represent the points and lines in another model. The results of Chapter 11 serve to establish the categoricalness of our system of real projective geometry. For they provide a definite naming of all the points and lines by sets of three real numbers, and this naming can be carried over into each model.

The whole problem of consistency and categoricalness, however, is connected with very difficult and deep questions, which have lately been investigated by philosophers and logicians, notably Gödel. Any adequate discussion would be beyond the scope of this book.

EXERCISE

Consider the following model for the geometry defined by the axioms of incidence (2·21–2·25) alone. *Points* are the 13 symbols $A_0, A_1, \ldots, A_{12}$; *lines* are the 13 symbols $a_0, a_1, \ldots, a_{12}$; A_i and a_j are incident if

$$i + j \equiv 0, 1, 3 \text{ or } 9 \pmod{13}.$$

Deduce that the 'geometry of incidence' is not categorical (Veblen).

12·2   Analytic geometry. We have remarked that the most satisfactory way to establish the logical consistency of our axioms, without
taking any geometrical ideas for granted, is by means of an algebraic
model. Such a model will now be described in detail.

A *point* is defined as an ordered set of three real numbers (x_1, x_2, x_3)
not all zero, with the understanding that $(\lambda x_1, \lambda x_2, \lambda x_3)$ is the same
point for any non-zero λ. Likewise a *line* is an ordered set of three
real numbers $[X_1, X_2, X_3]$, not all zero, with the understanding that
$[\lambda X_1, \lambda X_2, \lambda X_3]$ is the same line. For brevity we speak of the point (x)
and the line $[X]$. This point and line are said to be *incident* (the point
lying on the line and the line passing through the point) if and only if

12·21 $$\{xX\} = 0,$$

where

$$\{xX\} = x_1 X_1 + x_2 X_2 + x_3 X_3 = \Sigma x_i X_i.$$

Any figure or argument can be dualized by interchanging small and
capital letters, round and square brackets.

If (x) is a variable point on a fixed line $[X]$, we call 12·21 the *equation*
of the line $[X]$, meaning that it is the necessary and sufficient condition
for (x) to lie on $[X]$. Dually, if $[X]$ is a variable line through a fixed
point (x), we call the same relation the *equation* of the point (x), meaning that it is the condiition for $[X]$ to pass through (x). Thus the coordinates of a line or point are the coefficients in its equation (see § 11·8).

The three points (1, 0, 0), (0, 1, 0), (0, 0, 1) and the three lines [1, 0, 0],
[0, 1, 0], [0, 0, 1] form a triangle called the *triangle of reference* (see Fig.
12·2A). The point (1, 1, 1) and line [1, 1, 1] are called the *unit point* and

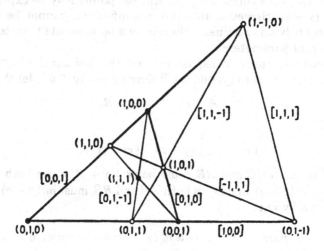

Fig. 12·2A

unit line. We shall see, in § 12·5, that there is nothing geometrically special about this triangle and point and line, apart from the fact that the point and line are trillinear pole and polar with respect to the triangle.

By eliminating $X_1 : X_2 : X_3$ from the three equations

$$\{xX\} = 0, \quad \{yX\} = 0, \quad \{zX\} = 0,$$

we find that the necessary and sufficient condition for three points (x), (y), (z) to be collinear is

12·22
$$\begin{vmatrix} x_1 & x_2 & x_3 \\ y_1 & y_2 & y_3 \\ z_1 & z_2 & z_3 \end{vmatrix} = 0.$$

This condition is equivalent to the existence of numbers λ, μ, ν, not all zero, such that

$$\lambda x_i + \mu y_i + \nu z_i = 0 \quad (i = 1, 2, 3).$$

If (y) and (z) are distinct points, $\lambda \neq 0$. Hence the general point collinear with (y) and (z) is $(\mu y_1 + \nu z_1, \mu y_2 + \nu z_2, \mu y_3 + \nu z_3)$ or, briefly,

$$(\mu y + \nu z).$$

When $\mu = 0$, this is the point (z) itself. For any other position, we can allow the coordinates of (y) to absorb the μ, and the collinear point is simply

$$(y + \nu z).$$

If we are concerned with only one such point, we may allow the ν to be absorbed too; thus three distinct collinear points may be expressed as (y), (z), $(y + z)$. However, this last simplification cannot be effected simultaneously on two lines if thereby one point would have to absorb two different parameters.

To illustrate these ideas, let us find the harmonic conjugate of $(y + z)$ with respect to (y) and (z). Referring to Fig. 2·5A, let the points

$$A, \quad B, \quad C, \quad Q, \quad R$$

be

$$(y), \quad (z), \quad (y + z), \quad (x), \quad (x + y).$$

Then P, on both QC and RB, must be $(x + y + z)$; S, on both QB and PA, must be $(x + z)$; and D, on both AB and RS, must be $(y - z)$. Hence, replacing z_i by νz_i:

12·23 *The harmonic conjugate of $(y + \nu z)$ with respect to (y) and (z) is* $(y - \nu z)$.

Since this result is independent of (x), we have here an analytic proof of 2·51.

Dually, the condition for three lines $[X]$, $[Y]$, $[Z]$ to be concurrent is

12·24
$$\begin{vmatrix} X_1 & X_2 & X_3 \\ Y_1 & Y_2 & Y_3 \\ Z_1 & Z_2 & Z_3 \end{vmatrix} = 0.$$

The general line concurrent with $[Y]$ and $[Z]$ is $[\mu Y + \nu Z]$, any such line except $[Z]$ can be expresses as

$$[Y + \nu Z],$$

and its harmonic conjugate with respect to $[Y]$ and $[Z]$ is $[Y - \nu Z]$.

EXERCISES

1. Show that the line joining $(1, 0, 0)$ to (x_1, x_2, x_3) is $[0, x_3, -x_2]$. Where does it meet $[1, 0, 0]$?

2. Name three lines through the point $(-1, 0, 1)$. Find their points of intersection with $[0, 0, 1]$.

3. If the triangle of reference is the diagonal triangle of a quadrangle having $(1, 1, 1)$ for one vertex, where are the other three vertices? (Cf. 2·42.)

4.* Show that the lines $[X]$ and $(y)(z)$ meet in the point

$$(\{zX\}y - \{yX\}z).$$

(*Hint:* What is the condition for $(y + \nu z)$ to lie on $[X]$?)

12·3 Verifying the axioms of incidence. To show that this analytic geometry really forms a model for the synthetic geometry developed in Chapter 2, we must verify that Axioms 2·21–2·25 are all satisfied.

The first four are easy. The point $(1, 0, 0)$ and line $[1, 0, 0]$ are certainly not incident. A line $[X_1, X_2, X_3]$ with $X_1 X_2 X_3 \neq 0$ is incident with three points such as

$$(0, X_3, -X_2), \quad (-X_3, 0, X_1), \quad (X_2, -X_1, 0);$$

$[0, X_2, X_3]$ is incident with

$$(0, X_3, -X_2), \quad (1, X_3, -X_2), \quad (1, 0, 0);$$

and so on. Two points (y) and (z) are incident with the unique line 12·22 or

---

* Graustein 1930, p. 70, Ex. 5. The idea of using capital letters for line coordinates is due to G.T. Bennett.

$$\left[\begin{vmatrix} y_2 & y_3 \\ z_2 & z_3 \end{vmatrix}, \begin{vmatrix} y_3 & y_1 \\ z_3 & z_1 \end{vmatrix}, \begin{vmatrix} y_1 & y_2 \\ z_1 & z_2 \end{vmatrix}\right],$$

and two lines $[Y]$ and $[Z]$ are incident with the point 12·24 or

$$\left(\begin{vmatrix} Y_2 & Y_3 \\ Z_2 & Z_3 \end{vmatrix}, \begin{vmatrix} Y_3 & Y_1 \\ Z_3 & Z_1 \end{vmatrix}, \begin{vmatrix} Y_1 & Y_2 \\ Z_1 & Z_2 \end{vmatrix}\right).$$

As for Desargues's theorem (our 2·25), let P, Q, R, and the point of concurrence be (x), (y), (z), and (u). Then there is no loss of generality in taking P', Q', R' to be

$$(x + u), \quad (y + u), \quad (z + u).$$

The point $QR \cdot Q'R'$, being collinear with (y) and (z) and also with $(y + u)$ and $(z + u)$, can only be $(y - z)$; similarly $RP \cdot R'P'$ and $PQ \cdot P'Q'$ are $(z - x)$ and $(x - y)$. The collinearity of these three points follows from the identity

$$(y_i - z_i) + (z_i - x_i) + (x_i - y_i) = 0.$$

EXERCISES

1. Find coordinates for the point of intersection of $[0, 1, -1]$ and $[-1, 1, 1]$ (Fig. 12·2A).

2. Prove Desargues's theorem as applied to the triangle of reference and $(k_1, 1, 1)(1, k_2, 1)(1, 1, k_3)$.

3. Work out §4·3, Ex. 3, taking $A_1 A_2 A_3$ to be the triangle of reference while $B_1 B_3 B_2$ is $(k_1, 1, 1)(1, k_2, 1)(1, 1, k_3)$. Obtain $k_1 k_2 k_3 = 1$ as the condition for lines $[0, -1, k_2]$, $[k_3, 0, -1]$, $[-1, k_1, 0]$ to be concurrent.

4. What further condition is required in Ex. 3 if also the lines $A_1 B_1$, $A_2 B_2$, $A_3 B_3$ are concurrent (so that the two triangles are 'quadruply perspective')?

12·4   Verifying the axioms of order and continuity. To show that this analytic geometry suffices for a model of the real projective geometry of Chapter 3, we still have to verify that Axioms 3·11–3·16 and 10·11 are satisfied when separation is suitably defined. The definition we shall adopt is as follows. Denoting points A, B, C, D by (a), (b), (c), (d), where

$$c_i = a_i + \mu b_i \quad \text{and} \quad d_i = a_i + \nu b_i,$$

we say that $AB//CD$ if and only if

$$\frac{\nu}{\mu} < 0$$

(so that μ and ν have opposite signs). We saw, in §12·2, that any third point collinear with two distinct points (a) and (b) can be expressed as $(a + \nu b)$ where $\nu \neq 0$; hence the above expressions for c_i and d_i merely mean that C and D are collinear with A and B.

Possibly the criterion $\nu/\mu < 0$ seems arbitrary; but it is really forced upon us by Pieri's definition for a segment (§3·6). According to that definition, the segment (ACB) is the locus of the harmonic conjugate of C with respect to two points $(a \pm \lambda b)$, where λ takes all values except 0. Since C is (c), where

$$c_i = a_i + \mu b_i = \frac{(\lambda + \mu)(a_i + \lambda b_i) + (\lambda - \mu)(a_i - \lambda b_i)}{2\lambda},$$

its harmonic conjugate (x) is given by

$$x_i = (\lambda + \mu)(a_i + \lambda b_i) - (\lambda - \mu)(a_i - \lambda b_i) = 2\mu\left(a_i + \frac{\lambda^2}{\mu}b_i\right),$$

i.e. the harmonic conjugate is $\left(a + \dfrac{\lambda^2}{\mu}b\right)$, where λ varies while μ is fixed. Since λ is real, the coefficient λ^2/μ takes in turn every value having the *same* sign as μ. Hence the supplementary segment AB/C consists of all points $(a + \nu b)$ for which ν has the *opposite* sign.

Axioms 3·11–3·13 can be verified immediately. To test 3·14, we observe that

$$a_i = c_i - \mu b_i, \quad d_i = c_i - (\mu - \nu)b_i,$$

$$\mu b_i = c_i - a_i, \quad \frac{\mu}{\nu}d_i = c_i - \left(1 - \frac{\mu}{\nu}\right)a_i.$$

Thus the three relations* $CB//AD$, $CA//BD$, $AB//CD$ mean

$$1 - \frac{\nu}{\mu} < 0, \quad 1 - \frac{\mu}{\nu} < 0, \quad \frac{\nu}{\mu} < 0,$$

one of which must hold whenever $\mu\nu(\mu - \nu) \neq 0$.

As for 3·15, the relations $AB//CD$ and $CA//BE$ mean that C, D, E are $(a + \mu b)$, $(a + \nu b)$, $(c + \rho a)$, where $\nu/\mu < 0$ and $-\rho < 0$. Since E is $(a + \mu b + \rho a)$ or $\left(a + \dfrac{\mu}{1 + \rho}b\right)$, the relation $AB//DE$ means that

$$\frac{\mu}{(1 + \rho)\nu} < 0,$$

which is obviously true if $\mu/\nu < 0$ and $\rho > 0$.

To test 3·16, suppose $ABCD \overset{S}{\barwedge} A'B'C'D'$. We may take S, A', B' to be

---

* Since $c_i/\mu = b_i + a_i/\mu$ and $d_i/\nu = b_i + a_i/\nu$, the relation $AB//CD$ is equivalent to $BA//CD$. This justifies our use of $CB//AD$ in place of $BC//AD$.

(s), $(a + \kappa s)$, $(b + \lambda s)$, and deduce

$$C' = A'B' \cdot CS = (a + \kappa s + \mu \overline{b + \lambda s}),$$

$$D' = A'B' \cdot DS = (a + \kappa s + \nu \overline{b + \lambda s}),$$

so that the relation $A'B'//C'D'$ means $\nu/\mu < 0$ again.

According to the definition in § 10·1, a sequence of collinear points

$$A_0 = (z), \quad A_1 = (y), \quad A_n = (y + \nu_n z) \quad (n = 2, 3, \ldots)$$

is monotonic if $A_0 A_n // A_1 A_{n+1}$ (for every $n > 1$). Since

$$A_1 = \left(z - \frac{y + \nu_n z}{\nu_n} \right) \quad \text{and} \quad A_{n+1} = \left(z + \frac{y + \nu_n z}{\nu_{n+1} - \nu_n} \right),$$

this condition amounts to $\nu_n/(\nu_{n+1} - \nu_n) > 0$, or

$$\frac{\nu_{n+1}}{\nu_n} > 1 \quad (n = 2, 3, \ldots),$$

which means that we have a monotonic sequence of numbers

$$\nu_1 = 0, \nu_2, \nu_3, \ldots .$$

We know from analysis that such a sequence of numbers is either convergent (to some limit ν) or divergent. In either case the sequence of points has a limit: $(y + \nu z)$ or (z), respectively.

We have now completed the identification of our synthetic and analytic geometries. A few further remarks arise naturally at this stage. Comparing the above results (e.g. 12·23) with § 11·5, we can identify the parameter ν with the abscissa of the point $(y + \nu z)$, referred to fundamental points

$$P_0 = (y), \quad P_1 = (y + z), \quad P_\infty = (z).$$

If ABC and $A'B'C'$ are any two sets of three collinear points, we may write

$$A = (a), \quad B = (b), \quad C = (a + b),$$

$$A' = (a'), \quad B' = (b'), \quad C' = (a' + b').$$

Then the analytic verification of the fundamental theorem 4·21 consists in the observation that if $ABCD \barwedge A'B'C'D'$, the abscissae of D and D' agree—

$$D = (a + \nu b) \quad \text{and} \quad D' = (a' + \nu b').$$

This parameter ν is called the *cross-ratio* of the four collinear points. In the notation of Veblen and Young it is $R(AB, DC)$.

More generally, if A, B, C, D are

$$(a), \quad (b), \quad (a + \mu b), \quad (a + \nu b),$$

we have $R(AB, DC) = v/\mu$, or $R(AB, CD) = \mu/v$. Thus the relation $ABCD \barwedge A'B'C'D'$ is equivalent to $R(AB, CD) = R(A'B', C'D')$,

$$AB//CD \text{ is equivalent to } R(AB, CD) < 0,$$

and

$$H(AB, CD) \text{ is equivalent to } R(AB, CD) = -1.$$

Let $[X]$ be any line through C, and $[Y]$ any line through D. Then

$$\{aX\} + \mu\{bX\} = 0, \quad \{aY\} + v\{bY\} = 0,$$

and therefore

$$\frac{\mu}{v} = \frac{\{aX\}\{bY\}}{\{bX\}\{aY\}}.$$

Following Heffter and Koehler,* let us call this the *cross-ratio* of the two points (a), (b) and the two lines $[X]$, $[Y]$. In particular, the two points are separated by the two lines if and only if the cross-ratio is negative.

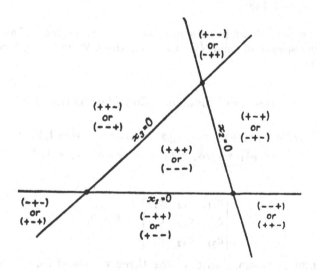

Fig. 12·4A

One simple consequence of this theory is the analytic interpretation of 3·81 as applied to the triangle of reference. If (a) and (b) are two points such that $a_1 a_2 b_1 b_2 \neq 0$, their cross-ratio with the sides $[1, 0, 0]$ and $[0, 1, 0]$ is

---

12·41
$$\frac{\{aX\}\{bY\}}{\{bX\}\{aY\}} = \frac{a_1 b_2}{b_1 a_2} = \frac{a_1}{a_2}\bigg/\frac{b_1}{b_2}.$$

Hence the two angular regions bounded by the lines $x_1 = 0$ and $x_2 = 0$ are distinguished by the sign of x_1/x_2 or of $x_1 x_2$; and the four triangular regions determined by the three lines $x_i = 0$ are distinguished by the signs of $x_2 x_3$, $x_3 x_1$, $x_1 x_2$. In the 'interior' region containing $(1, 1, 1)$ the three coordinates x_i all have the same sign; but in the remaining three regions one coordinate differs in sign from the other two (see Fig. 12·4A).

A similar distinction can be made as to the signs of the coordinates of a line. If $\{xX\} = 0$, the three products $x_i X_i$ cannot all have the same sign. Hence any line for which $X_1 X_2 X_3 \neq 0$ must be exterior to the region where the point coordinates have the same distribution of signs as these line coordinates.

EXERCISES

1. Show that the three relations $AB//CD$, $AB//CE$, $AB//DE$ cannot all hold simultaneously (cf. 3·18).

2. Prove in detail that if $\{v_n\}$ is a divergent monotonic sequence of numbers, the limit of the sequence of points $(y + v_n z)$ is (z). (*Hint:* Write $(y + v_n z)$ in the form $(z + v_n^{-1}y)$.)

12·5 The general collineation. Consider the transformation

12·51
$$\begin{cases} x_i' = c_{i1}x_1 + c_{i2}x_2 + c_{i3}x_3 = \Sigma c_{ij}x_j & (i = 1, 2, 3), \\ X_j = c_{1j}X_1' + c_{2j}X_2' + c_{3j}X_3' = \Sigma c_{ij}X_i' & (j = 1, 2, 3), \end{cases}$$

where

$$\begin{vmatrix} c_{11} & c_{12} & c_{13} \\ c_{21} & c_{22} & c_{23} \\ c_{31} & c_{32} & c_{33} \end{vmatrix} = \Delta \neq 0.$$

(The Σ implies summation over the three values of i or j, whichever of these letters appears *twice* in the expression.) This transformation leads from a point (x) to a point (x') and from a line $[X]$ to a line $[X']$. Since

$$\{x'X'\} = \Sigma x_i'X_i' = \Sigma\Sigma c_{ij}x_jX_i' = \Sigma x_jX_j = \{xX\},$$

it preserves incidence; and since $\Delta \neq 0$, it is one-to-one. Hence it is a collineation, as defined in §4·1.

Solving the equations 12·51 for x_j and X_i', we obtain the following expressions for $[X']$ in terms of $[X]$ and for (x) in terms of (x'):

12·52 $\begin{cases} X_i' = C_{i1}X_1 + C_{i2}X_2 + C_{i3}X_3 = \Sigma C_{ij}X_j & (i = 1, 2, 3), \\ x_j = C_{1j}x_1' + C_{2j}x_2' + C_{3j}x_3' = \Sigma C_{ij}x_i' & (j = 1, 2, 3), \end{cases}$

where C_{ij} is the cofactor of c_{ij} divided* by the determinant Δ, so that

$$\Sigma c_{ij}C_{ik} = \delta_{jk},$$

which means 1 or 0 according as $j = k$ or $j \neq k$ (the 'Kronecker delta'). These solutions may be verified as follows:

$$\Sigma C_{ij}X_j = \Sigma\Sigma C_{ij}c_{kj}X_k' = \Sigma\delta_{ik}X_k' = X_i',$$

$$\Sigma C_{ij}x_i' = \Sigma\Sigma C_{ij}c_{ik}x_k = \Sigma\delta_{jk}x_k = x_j.$$

Given a triangle (a) (b) (c), we may describe the position of any point P by means of *barycentric* coordinates, defined as follows: If P does not coincide with the vertex (a), it can be joined to (a) by a definite line that meets the opposite side (b) (c) in a point $(\mu b + \nu c)$. Then P, being collinear with (a) and $(\mu b + \nu c)$, may be expressed as

$$(\lambda a + \mu b + \nu c).$$

The barycentric coordinates are these coefficients λ, μ, ν. The point (a) itself is included by allowing both μ and ν to vanish. By absorption we may take any particular point not on a side of the triangle to have $\lambda = \mu = \nu = 1$.

When (a) (b) (c) and $(a + b + c)$ are the triangle of reference and unit point, $(\lambda a + \mu b + \nu c)$ is (λ, μ, ν) and the barycentric coordinates are the same as the ordinary coordinates. The collineation

12·53 $$x_i' = a_i x_1 + b_i x_2 + c_i x_3,$$

transforming the quadrangle $(1, 0, 0)$ $(0, 1, 0)$ $(0, 0, 1)$ $(1, 1, 1)$ into $(a)(b)(c)(a + b + c)$, transforms (λ, μ, ν) into the point

$$(\lambda a + \mu b + \nu c),$$

which has these same barycentric coordinates referred to the new quadrangle instead of the old.

Since *any* quadrangle can be written as $(a)(b)(c)(a + b + c)$, we see from 5·12 that 12·51 or 12·53 is the *most general* collineation. The condition $\Delta \neq 0$ or

$$\begin{vmatrix} a_1 & b_1 & c_1 \\ a_2 & b_2 & c_2 \\ a_3 & b_3 & c_3 \end{vmatrix} \neq 0$$

merely expresses the requirement that the points (a), (b), (c) form a

---

* Since the coordinates are homogeneous, there would be no harm in defining C_{ij} to be just the cofactor of c_{ij}, without troubling to divide by Δ.

triangle. Thus the equations 12·51 may be regarded *either* as a collineation shifting the points in the plane *or* as a coordinate transformation
giving a new name to each point.

Defining

$$A = \{aX\}, \quad B = \{bX\}, \quad C = \{cX\},$$

so that the points (a), (b), (c) have equations $A = 0$, $B = 0$, $C = 0$, we
find that the point $(\lambda a + \mu b + \nu c)$ has the equation

$$\lambda A + \mu B + \nu C = 0.$$

Thus the barycentric coordinates of any point are simply the coefficients of A, B, C when its equation is expressed in terms of the equations of those points. The above remarks serve to justify Möbius's
'barycentric calculus' (so effectively used by Baker), where the general
point is denoted by

$$\lambda A + \mu B + \nu C$$

(with '$= 0$' omitted). In this notation our triangle of reference (formed
by the points $X_1 = 0$, $X_2 = 0$, $X_3 = 0$) is simply $X_1 X_2 X_3$, and the point (x)
is $x_1 X_1 + x_2 X_2 + x_3 X_3$. Thus

$$(x_1, x_2, x_3) = x_1(1, 0, 0) + x_2(0, 1, 0) + x_3(0, 0, 1),$$

as in vector analysis. (In fact, if we think of the plane as lying in an
affine space, we may interpret these symbols as vectors leading from
some fixed origin outside the plane to the points considered; then the
expression $\lambda A + \mu B + \nu C$ is a sum of vectors.) Historically, this barycentric calculus (1827) preceded Plücker's line coordinates (1828–30).
Von Staudt's synthetic approach to projective geometry came later
still, as we have seen. Grassmann, a contemporary of von Staudt,
developed a 'calculus of extension' in which both points and lines are
represented as vectors: the vector product of two points is their join,
and the vector product of two lines is their intersection (cf. § 12·3).

One very practical rule emerges from this little digression. When
seeking an analytic proof for a theorem concerning a triangle, we are
justified in taking this as triangle of reference; and any fixed point not
on a side of the triangle may be named (1, 1, 1). Thus Ex. 2 of § 12·3
would suffice for a proof of Desargues's theorem (our 2·25), and Ex. 3 for
a proof of Pappus's theorem (4·31). (This is far neater than the proofs of
Pappus's theorem given in most textbooks on analytic geometry.) For
theorems involving a quadrangle it is often convenient to take the
vertices to be $(1, \pm 1, \pm 1)$, so that the six sides are $x_i \pm x_j = 0$ $(i < j)$
and the diagonal triangle is the triangle of reference. Dually, a given
quadrilateral may be taken to have sides $[1, \pm 1, \pm 1]$ and vertices
$X_i \pm X_j = 0$.

The following special collineations will be found useful: a homology with centre $(0, 0, 1)$ and axis $[0, 0, 1]$,

12·54
$$x_1' = x_1, \quad x_2' = x_2, \quad x_3' = \frac{x_3}{\rho};$$

and an elation with centre $(c_1, c_2, 0)$ and axis $[0, 0, 1]$,

12·55 $\qquad x_1' = x_1 + c_1 x_3, \quad x_2' = x_2 + c_2 x_3, \quad x_3' = x_3.$

EXERCISES

1. Given an analytic proof for §2·4, Ex. 1.

2. Find the collineations that transform $(1, 0, 0)$ $(0, 1, 0)$ $(0, 0, 1)$ $(1, 1, 1)$ into the following quadrangles:

        (i) $(1, 0, 0)$ $(0, 1, 0)$ $(0, 0, 1)$ (a_1, a_2, a_3),

        (ii) $(-1, 1, 1)$ $(1, -1, 1)$ $(1, 1, -1)$ $(1, 1, 1)$,

        (iii) $(0, 1, 0)$ $(0, 0, 1)$ $(1, 0, 0)$ $(1, 1, 1)$,

        (iv) $\quad (0, 1, 0)$ $(0, 0, 1)$ $(1, 1, 1)$ $(1, 0, 0)$.

The last two collineations are periodic. What are their periods?

3. Find the collineation that interchanges $(\pm 1, 1, 1)$ and also interchanges $(\pm 1, -1, 1)$ (cf. 5·31). Where are the centre and axis of this harmonic homology?

4. Given an analytic proof for the exercise to §5·3.

5. Find the elation with axis $[1, 0, 0]$ transforming $(1, 0, 0)$ into $(1, 1, 0)$ (cf. 5·22). Where is its centre?

6. Find the homology or elation that transforms the triangle of reference into $(k_1, 1, 1)$ $(1, k_2, 1)$ $(1, 1, k_3)$ (cf. 5·24). When will it be an elation? (*Hint:* Write $K_i = 1/(k_i - 1)$.)

7. Express the following three collineations in terms of line coordinates. Find their invariant points and lines.

      (i) $x_1' = x_1, \qquad x_2' = x_3, \qquad\qquad x_3' = -x_2;$

      (ii) $x_1' = c_{11} x_1, \quad x_2' = x_2, \qquad\qquad x_3' = c_{32} x_2 + x_3;$

      (iii) $x_1' = x_1, \qquad x_2' = c_{21} x_1 + x_2, \quad x_3' = c_{31} x_1 + c_{32} x_2 + x_3.$

12·6 The general polarity. Since the product of any two correlations is a collineation, the general correlation can be obtained by combining the general collineation 12·51 with the special correlation that interchanges X_i' and x_i', thus:

$$\begin{cases} X_i' = \Sigma c_{ij} x_j & (i = 1, 2, 3), \\ X_j = \Sigma c_{ij} x_i' & (j = 1, 2, 3). \end{cases}$$

(All incidences are dualized, as $\Sigma x_i' X_i' = \Sigma X_j x_j$.)

This correlation is a *polarity* if it is equivalent to the inverse correlation $X_j' = \Sigma c_{ij} x_i$ or (interchanging i and j)

$$X_i' = \Sigma c_{ji} x_j \quad (i = 1, 2, 3).$$

This means that $c_{ji} = \lambda c_{ij}$, with the same λ for all i and j, so that $c_{ij} = \lambda c_{ji} = \lambda^2 c_{ij}$, $\lambda^2 = 1$, $\lambda = \pm 1$; but we cannot have $\lambda = -1$, as that would make

$$\Delta = \begin{vmatrix} 0 & c_{12} & -c_{31} \\ -c_{12} & 0 & c_{23} \\ c_{31} & -c_{23} & 0 \end{vmatrix} = 0.$$

Hence $\lambda = 1$ and $c_{ji} = c_{ij}$. In other words, a correlation is a polarity if and only if the matrix of coefficients c_{ij} is symmetric.

To emphasize this extra condition we shall write a_{ij} ($= a_{ji}$) instead of c_{ij}. Moreover, the nature of a polarity is such that no confusion can be caused by omitting the prime ['] and writing simply

12·61 $X_i = \Sigma a_{ij} x_j \quad (i = 1, 2, 3).$

These equations give us the polar $[X]$ of a given point (x). Solving them, we obtain the pole (x) of a given line $[X]$ in the form

12·62 $x_i = \Sigma A_{ij} X_j \quad (i = 1, 2, 3),$

and we know that the coefficients are connected as follows:

$$a_{ji} = a_{ij}, \quad A_{ji} = A_{ij}, \quad \Sigma a_{ij} A_{ik} = \delta_{jk},$$

so that

$$\det(a_{ij}) = \Delta \neq 0, \quad \det(A_{ij}) = \Delta^{-1}.$$

Two points (x) and (y) are *conjugate* (§5·5) if (x) lies on the polar $[Y]$ of (y). Since $Y_i = \Sigma a_{ij} y_j$, the condition $\{xY\} = 0$ or $\Sigma x_i Y_i = 0$ becomes

$$\Sigma\Sigma a_{ij} x_i y_j = 0.$$

Letting (x) vary, we see that this is *the equation for the polar of* (y). We shall often write it in the abbreviated form

12·63 $(xy) = 0.$

Dually, the condition for lines $[X]$ and $[Y]$ to be conjugate, or the equation for the pole of $[Y]$, is

12·64 $[XY] = 0,$

where $[XY] = \Sigma\Sigma A_{ij} X_i Y_j$.

The fact that a polarity induces an *involution* of conjugate points on any non-self-conjugate line (5·53) may be verified by writing down the condition for points $(x, 1, 0)$ and $(x', 1, 0)$ (on the fixed line $[0, 0, 1]$) to be conjugate, viz.

$$a_{11}xx' + a_{12}(x + x') + a_{22} = 0.$$

As we saw, just before 11·43, this is a proper involution unless

$$a_{11}a_{22} - a_{12}^2 = 0,$$

in which case $A_{33} = 0$, and the line $[0, 0, 1]$ is self-conjugate.

For an analytic proof of Chasles's theorem (5·61), we apply the general polarity 12·61 to the vertices of the triangle of reference, obtaining the sides

$$[a_{11}, a_{21}, a_{31}], \quad [a_{12}, a_{22}, a_{32}], \quad [a_{13}, a_{23}, a_{33}]$$

of another triangle. Since the result is trivial when a pair of corresponding sides coincide, we may assume that at least two of a_{23}, a_{31}, a_{12} are different from zero. Then any two corresponding sides are concurrent with

$$[a_{31}a_{12}, a_{12}a_{23}, a_{23}a_{31}].$$

For von Staudt's converse theorem (5·71), we observe that the sides of the triangle of reference are related to the points

$$(k_1, 1, 1), \quad (1, k_2, 1), \quad (1, 1, k_3)$$

by the polarity 12·62 with $A_{ii} = k_i$ and every other $A_{ij} = 1$, namely,

$$x_1 = k_1 X_1 + X_2 + X_3,$$
$$x_2 = X_1 + k_2 X_2 + X_3,$$
$$x_3 = X_1 + X_2 + k_3 X_3$$

(see § 12·3, Ex. 2).

Returning to the polarity 12·61 or 12·63, we observe that the condition for $(0, 1, 0)$ and $(0, 0, 1)$ to be conjugate is $a_{23} = 0$. Thus the triangle of reference is self-polar if and only if

$$a_{23} = a_{31} = a_{12} = 0.$$

By choosing such a triangle of reference we reduce a polarity to its *canonical form*

$$X_i = a_{ii}x_i \quad (i = 1, 2, 3; \quad a_{11}a_{22}a_{33} = \Delta \neq 0).$$

The coefficients a_{ii} are determined by one further pole and polar, as in 5·63. In fact, if ABC is the triangle of reference while P is (C_1, C_2, C_3) and p is $[c_1, c_2, c_3]$, then the polarity $(ABC)(Pp)$ is of the above form with $a_{ii} = c_i/C_i$. In particular, the canonical polarity relating $(1, 1, 1)$ to $[c_1, c_2, c_3]$ is

12·65 $X_i = c_i x_i \quad (i = 1, 2, 3; \quad c_1 c_2 c_3 \neq 0).$

So far, we have insisted that the determinant Δ shall not vanish. It is interesting to see what kind of degenerate polarity remains if we allow $\Delta = 0$. The relations 12·61 still provide a unique line $[X]$ for each point (x), but now all polars $[X]$ pass through one fixed point, and each is the polar of infinitely many (x)'s. In fact, the vanishing of the determinant implies the existence of numbers z_1, z_2, z_3, not all zero, such that

$$\Sigma a_{ij} z_i = 0 \quad (j = 1, 2, 3).$$

Hence, for any point (x), $\Sigma\Sigma a_{ij} z_i x_j = 0$; which means that the polar $[X]$, satisfying $\Sigma z_i X_i = 0$, always passes through a certain point (z), which is the A of §5·9. Such a line $[X]$ is also the polar of $(x + vz)$ for any v.

In other words, when $\Delta = 0$, there is a point (z) that is conjugate to every point (x). This universal conjugate point (z) is unique unless all points have the same polar $[X]$. This doubly degenerate case arises when a_{ij} is of the form $a_i a_j$, so that 12·61 reduces to

$$X_i = a_i \Sigma a_j x_j.$$

This means that all cofactors of order 2 in Δ vanish, so that the matrix (a_{ij}) is of rank 1.

If (a_{ij}) is of rank 2, so that (z) is unique, the condition for two points (x) and (y) to be conjugate is still 12·63 and there is still an involution of conjugate points on any line not passing through (z). Of such a pair of points, each is joined to (z) by the polar of the other; thus we have an involution of conjugate lines through (z). For each pair of lines in this involution the polarity relates every point on either line to the other line. In this sense, the degenerate polarity *is* the involution of conjugate lines through (z).

Dually, the transformation 12·62 with $\det(A_{ij}) = 0$ represents the second kind of degenerate polarity, such that the pole of any line lines on one fixed line $[Z]$. If the matrix (A_{ij}) is of rank 1, we have the second kind of double degeneracy—all lines have the same pole; but if it is of rank 2, the line $[Z]$ is unique and the degenerate polarity is essentially an involution of point pairs on $[Z]$.

The above remarks may be summarised as follows:

12·66   *If (a_{ij}) is a matrix of rank 2, the polarity $X_i = \Sigma a_{ij} x_j$ degenerates into an involution of line pairs through a fixed point. Dually, if (A_{ij}) is of rank 2, the polarity $x_i = \Sigma A_{ij} X_j$ degenerates into an involution of point pairs on a fixed line.*

EXERCISES

1. Prove Hesse's theorem (our 5·55), using the general polarity and the quadrilateral $[1, \pm 1, \pm 1]$, whose vertices are

$$(0, 1, \pm 1), \quad (\pm 1, 0, 1), \quad (1, \pm 1, 0).$$

2. Prove 5·62, using the triangle of reference.

3. Verify 6·21 and 6·22 as applied to the triangle of reference and the unit line $[1, 1, 1]$.

4. Solve Ex. 6 of §7·7, using the triangle of reference.

5. Use the following coordinates in §5·7, Ex. 2:

$$A(1, 0, 0), \qquad B(0, 1, 0), \qquad C'(0, 0, 1),$$

$$L(0, 1, p), \qquad M(q, 0, 1), \qquad C(1, r, 0),$$

$$B'N[0, 1, P], \quad A'N[Q, 0, 1], \quad A'B'[1, R, 0].$$

The incidences in Fig. 4·3A require

$$1 + qQ + (rR)^{-1} = 0, \quad 1 + rR + (pP)^{-1} = 0.$$

Verify that these equations imply $1 + pP + (qQ)^{-1} = 0$ (thus providing an alternative proof for Pappus's theorem) and $pqrPQR = 1$. Finally, obtain the condition $pqr = PQR = \pm 1$ for the duality suggested by the above notation to be induced by a polarity.

6. Verify that the relations

$$X_1 = \lambda c_1' x_1, \quad X_2 = c_2 x_2, \quad X_3 = (c_3 + \lambda c_3') x_3$$

define a pencil of polarities transforming the unit point $(1, 1, 1)$ into the pencil of lines concurrent with $[0, c_2, c_3]$ and $[c_1', 0, c_3']$ and that this is a self-dual system (i.e. a range as well as a pencil) if $c_3' = 0$. Setting $c_1' = 1$, we thus obtain the system

$$X_1 = \lambda x_1, \quad X_2 = c_2 x_2, \quad X_3 = c_3 x_3$$

(cf. 5·82). Verify that the locus of poles of $[1, 1, 1]$ is the line $c_2 x_2 = c_3 x_3$.

7. Find the locus of poles of a fixed line $[X]$ with respect to polarities

$$X_1 = x_3 + \lambda x_1, \quad X_2 = x_2, \quad X_3 = x_1.$$

(*Hint:* Rewrite these relations as $\rho X_1 = x_3 + \lambda x_1$, $\rho X_2 = x_2$, $\rho X_3 = x_1$; then eliminate ρ and λ.)

12·7 Conics. The condition for a point (x) to be self-conjugate for the polarity 12·61 is $(xx) = 0$, or

$$a_{11}x_1^2 + a_{22}x_2^2 + a_{33}x_3^2 + 2a_{23}x_2 x_3 + 2a_{31}x_3 x_1 + 2a_{12}x_1 x_2 = 0.$$

Hence the polarity is elliptic or hyperbolic according as the quadratic form (xx) (with determinant $\Delta \neq 0$) is definite or indefinite.* In the latter case the locus of self-conjugate points is the conic†

$$(xx) = 0,$$

and the envelope of self-conjugate lines is the same conic in the form

$$[XX] = 0.$$

In particular, the condition for (x) to be self-conjugate for 12·65 is $\Sigma c_i x_i^2 = 0$; thus the canonical polarity is elliptic or hyperbolic according as the three non-vanishing coefficients c_i do or do not have the same sign, and in the latter case the conic is

$$c_1 x_1^2 + c_2 x_2^2 + c_3 x_3^2 = 0 \quad \text{or} \quad \frac{X_1^2}{c_1} + \frac{X_2^2}{c_2} + \frac{X_3^2}{c_3} = 0.$$

By the coordinate transformation

$$x_i \to |c_i|^{-1/2} x_i, \quad X_i \to |c_i|^{1/2} X_i$$

we can reduce the coefficients to ± 1. Then, renumbering the three coordinates if necessary, 12·65 becomes

12·71 $$X_1 = x_1, \quad X_2 = x_2, \quad X_3 = \pm x_3$$

with the upper or lower sign according as the polarity is elliptic or hyperbolic. [In the former case this amounts to taking $(1, 1, 1)$ to be one of the four points described in §7·7, Ex. 6.] Thus any non-degenerate conic may be expressed as

12·72 $$x_1^2 + x_2^2 - x_3^2 = 0, \quad X_1^2 + X_2^2 - X_3^2 = 0.$$

In this form, the triangle of reference is self-polar. Another useful equation,

12·73 $$x_2^2 - x_3 x_1 = 0$$

(where the triangle of reference is formed by two tangents and the join of their points of contact), is derived from 12·72 by the transformation

$$x_1 \to \tfrac{1}{2}(x_1 - x_3), \quad x_2 \to x_2, \quad x_3 \to \tfrac{1}{2}(x_1 + x_3).$$

This exhibits the conic as the locus of the point of intersection of the projectively related lines $x_1 - t x_2 = 0$ and $x_2 - t x_3 = 0$, as in 6·54. In other words, the conic is the locus of the point $(t^2, t, 1)$ whose *parameter* is t.

---

* A necessary and sufficient condition for a *definite* form (or *elliptic* polarity) is that the three numbers a_{11}, A_{22}, Δ all have the same sign. See Veblen and Young 1918, p. 205.
† Hesse 1897 seems to have been the first to write the equation for a conic in the form $\Sigma\Sigma a_{ij} x_i x_j = 0$.

Since the condition for the conic $(xx) = 0$ to pass through $(1, 0, 0)$ is $a_{11} = 0$, the general conic circumscribing the triangle of reference is

$$a_{23}x_2x_3 + a_{31}x_3x_1 + a_{12}x_1x_2 = 0 \quad (a_{23}a_{31}a_{12} \neq 0).$$

The coordinate transformation

$$x_1 \to a_{23}x_1, \quad x_2 \to a_{31}x_2, \quad x_3 \to a_{12}x_3$$

converts this into

12·74 $$x_2x_3 + x_3x_1 + x_1x_2 = 0$$

or $x_1^{-1} + x_2^{-1} + x_3^{-1} = 0$. This exhibits the conic as the locus of trilinear poles of lines through the unit point $X_1 + X_2 + X_3 = 0$ (which is the point described in §7·7, Ex. 5). Working out the cofactors in the determinant, we obtain the envelope equation

$$X_1^2 + X_2^2 + X_3^2 - 2X_2X_3 - 2X_3X_1 - 2X_1X_2 = 0$$

or $X_1^{1/2} \pm X_2^{1/2} \pm X_3^{1/2} = 0$. Dually, a conic inscribed in the triangle of reference is

12·75 $\quad X_2X_3 + X_3X_1 + X_1X_2 = 0 \quad$ or $\quad x_1^{1/2} \pm x_2^{1/2} \pm x_3^{1/2} = 0$.

If $\Delta = 0$, the equation $(xx) = 0$ represents a *degenerate conic*, the locus of self-conjugate points for a degenerate polarity (see pp. 72, 86) in which every point (x) is conjugate to one special point (z). If (y) is self-conjugate, every point collinear with (y) and (z) must be likewise self-conjugate, since

$$\Sigma\Sigma a_{ij}(y_i + vz_i)(y_j + vz_j) = (yy) + 2v(yz) + v^2(zz) = 0.$$

In the doubly degenerate case when $a_{ij} = a_ia_j$, the equation $(xx) = 0$ reduces to $\{ax\}^2 = 0$, which is essentially the line $[a]$. Thus the equation $(xx) = 0$ with $\det(a_{ij}) = 0$ represents a degenerate conic locus consisting of a single point [for example, $x_1^2 + x_2^2 = 0$, which is $(0, 0, 1)$] or a line (for example, $x_1^2 = 0$) or two lines (for example, $x_1x_2 = 0$); but not more than two, or the equation would involve the product of three or more linear factors.

Dually, the equation $[XX] = 0$ with $\det(A_{ij}) = 0$ represents a degenerate conic envelope consisting of a line (for example, $X_1^2 + X_2^2 = 0$, which is $[0, 0, 1]$) or a point (for example, $X_1^2 = 0$) or two points (for example, $X_1X_2 = 0$).*

The following result is typical of many applications of the (xy) notation. The condition for the join of two points (x) and (y) to be a tangent to the conic $(xx) = 0$ is

---

* The tangents to a very small circle visibly resemble a pencil of lines, enveloping a single point; and the tangents to a very flat ellipse (with eccentricity nearly 1) resemble two such pencils. See Robson 1940, p. 67.

12·76 $(xx)(yy) - (xy)^2 = 0.$

To see this, let $(x + \mu y)$ be the point of contact of such a tangent. The point of contact must be conjugate to both (x) and (y); hence

$$(x\ x + \mu y) = 0, \qquad (x + \mu y\ y) = 0,$$

i.e.

$$(xx) + \mu(xy) = 0, \quad (xy) + \mu(yy) = 0.$$

We obtain 12·76 by eliminating μ; and the argument can be reversed. The same equation may be obtained as the degeneracy condition for the involution

$$(xx) + (xy)(\mu + \mu') + (yy)\mu\mu' = 0$$

of conjugate points $(x + \mu y)$ and $(x + \mu' y)$ on the line $(x)(y)$.

If (y) is a fixed exterior point, 12·76 is the combined equation for the two tangents that can be drawn to the conic from that point. Dually, the conic $[XX] = 0$ meets a secant $[Y]$ in the two points

$$[XX][YY] - [XY]^2 = 0.$$

For problems involving two conics, it is convenient to use the notation $(xx)' = \Sigma\Sigma a'_{ij} x_i x_j$. If two conics $(xx) = 0$ and $(xx)' = 0$ have four points of intersection, the equation

12·77 $(xx) + \lambda(xx)' = 0$

represents a conic (possibly degenerate) through the same four points. Moreover, this is the most general conic of the quadrangular pencil; for if (y) is any fifth point on such a conic, we merely have to choose λ so as to satisfy $(yy) + \lambda(yy)' = 0$. Taking a more general standpoint, let us call 12·77 a *pencil of conics* even if the conics have no common self-polar triangle (so that the definition implied in 6·82 cannot be used). We shall find that most of the theorems about pencils remain valid. Extending 5·81, we observe that the polars of (y) form the pencil of lines

$$(xy) + \lambda(xy)' = 0$$

whose centre, given by solving the two equations $(xy) = (xy)' = 0$ for (x), is conjugate to (y) with respect to every one of the conics. As for 6·81: the poles of a line $(y)(z)$ all satisfy the equation

12·78 $(xy)(xz)' - (xz)(xy)' = 0,$

which is obtained by eliminating λ from

$$(xy) + \lambda(xy)' = 0, \quad (xz) + \lambda(xz)' = 0.$$

The common conjugates of points $(\mu y + z)$ on the line $(y)(z)$ satisfy the same equation, obtained by eliminating μ from

$$\mu(xy) + (xz) = 0, \quad \mu(xy)' + (xz)' = 0.$$

Desargues's involution theorem (our 6·72 and 6·59) can be extended as follows: The conic 12·77 passes through $(\mu y + z)$ if

$$\mu^2(yy) + 2\mu(yz) + (zz) + \lambda[\mu^2(yy)' + 2\mu(yz)' + (zz)'] = 0.$$

By considering the sum and products of the roots of this quadratic equation for μ, we see that two points $(\mu_1 y + z)$ and $(\mu_2 y + z)$ lie on the same conic of the pencil if

$$(\mu_1 + \mu_2)[(yy) + \lambda(yy)'] = -2[(yz) + \lambda(yz)']$$

and

$$\mu_1\mu_2[(yy) + \lambda(yy)'] = (zz) + \lambda(zz)'.$$

Eliminating λ, we obtain the equation

$$2\mu_1\mu_2[(yy)(yz)' - (yz)(yy)'] + (\mu_1 + \mu_2)[(yy)(zz)' - (zz)(yy)']$$
$$+ 2[(yz)(zz)' - (zz)(yz)'] = 0,$$

which is also the condition for the points $(\mu_1 y + z)$ and $(\mu_2 y + z)$ to be conjugate with respect to 12·78. Hence:

Those conics of the pencil 12·77 that meet the line $(y)(z)$ do so in pairs of points that are conjugate with respect to the conic 12·78.

In the case of a quadrangular pencil, there are three values of λ for which 12·77 consists of a line pair. This fact is neatly employed in the following proof* of Pascal's theorem (our 7·21). Using the notation of Fig. 7·2A, let the lines BB', CA', and AC' be $[X]$, $[Y]$, and $[Z]$. Then for a certain λ the equation

$$(xx) + \lambda\{Xx\}\{Yx\} = 0$$

represents the line pair BA', CB'; for a certain μ the equation

$$(xx) + \mu\{Xx\}\{Zx\} = 0$$

represents the pair BC', AB'. Subtracting these, we obtain another degenerate conic

$$\{Xx\}(\lambda\{Yx\} - \mu\{Zx\}) = 0$$

through the common points B, B', N, L of the first two. Now, the first line pair is BN, LB' and the second is BL, NB'; hence the third must be BB', NL. The factor $\{Xx\}$ gives the line $[X]$ which is BB'; therefore, NL is $[\lambda Y - \mu Z]$, concurrent with CA' and AC'.

An interesting special case of 12·78 arises when the four common points are $(1, \pm 1, \pm 1)$, so that the quadrangular pencil of conics is

---

* Robson 1947, p. 91.

given by

12·79 $c_1 x_1^2 + c_2 x_2^2 + c_3 x_3^2 = 0, \quad c_1 + c_2 + c_3 = 0$

for various values of the c's. The polar of a fixed point (x) is

$$[c_1 x_1, c_2 x_2, c_3 x_3],$$

which continually passes through the fixed point $(x_1^{-1}, x_2^{-1}, x_3^{-1})$. This 'quadratic transformation' $(x) \to (x^{-1})$ (which somewhat resembles inversion with respect to a circle) transforms the points of a line $[X]$ into the points of a conic

$$X_1 x_1^{-1} + X_2 x_2^{-1} + X_3 x_3^{-1} = 0$$

through the vertices of the triangle of reference. This conic is also the locus of poles of $[X]$, as we see by eliminating the c's from

$$X_i = c_i x_i \quad \text{and} \quad \Sigma c_i = 0.$$

EXERCISES

1. Verify 6·43 as applied to the quadrangle $(1, \pm 1, \pm 1)$.

2. Show that a unique conic can be drawn through $(1, 1, 1)$ to touch $[0, 0, 1]$ at $(1, 0, 0)$ and $[1, 0, 0]$ at $(0, 0, 1)$ (cf. 6·53 and 12·73).

3. Verify §6·5, Ex. 3, taking the exterior points $p \cdot s$ and $q \cdot r$ to be (y) and (z). The conic through the six points turns out to be

$$(xx)(yz) - (xy)(xz) = 0.$$

4. If $[Z]$ meets $(xx) = 0$ in two points, prove that the two lines joining these points to another point (y) are given by

$$\{Zy\}^2 (xx) - 2\{Zx\}\{Zy\}(xy) + \{Zx\}^2 (yy) = 0.$$

(*Hint:* What value of μ will make $(x + \mu y)$ lie on the conic?)

5. If the sides of a variable triangle pass through three fixed points $(\lambda, 1, 0)$, $(1, \mu, 0)$, $(1, 1, \nu)$, while the vertices opposite to the first two sides run along the respective lines $[1, 0, 0]$, $[0, 1, 0]$, prove that the third vertex will trace a conic or a line (as in 6·61 or the Exercise to §4·2). (*Hint:* Take the triangle to be

$$(0, x_2 - \mu x_1, x_3)(x_1 - \lambda x_2, 0, x_3)(x_1, x_2, x_3).)$$

6. Show that the conics

$$(xx) + \lambda(xy)^2 = 0$$

form a self-dual system (p. 89, cf. 12·76). What happens when (y) is an interior point?

7. Considering the conic 12·73 as the locus of $(t^2, t, 1)$, prove that the secant joining the points with parameters t and t' is $[1, -(t + t'), tt']$ and that the tangent

at the point t is $[1, -2t, t^2]$. Deduce the envelope equation

$$X_2^2 - 4X_3 X_1 = 0$$

and check this by direct computation of cofactors.

8. Show that the quadratic transformation $(x) \to (x^{-1})$ transforms a conic through two vertices of the triangle of reference into a conic through the same two vertices. How many conics are transformed into themselves? (Six pencils.)

9. Give an analytic treatment of §7·2, Ex. 6, taking the conic in the form 12·74, with ABC for triangle of reference. (*Hint:* Take $A_1 B_1 C_1$ to be $(\lambda, 1, -1)(-1, \mu, 1)$ $(1, -1, \nu)$. The point of concurrence turns out to be (λ, μ, ν).)

12·8   The affine plane: affine and areal coordinates.

We saw, in §8·1, how the affine plane can be derived from the projective plane by removing one line. Analytically, the simplest way to do this is to remove one side of the triangle of reference, say $[0, 0, 1]$ or $x_3 = 0$. The remaining two sides are then called coordinate *axes*. Any point for which $x_3 \neq 0$ can be normalized (dividing through by x_3) so as to take the form $(x_1, x_2, 1)$, which can then be abbreviated to (x_1, x_2). In this manner we obtain a unique symbol for every ordinary point. These non-homogeneous coordinates x_1, x_2 are called *affine* coordinates.

Lines $[X_1, X_2, X_3]$, where X_1 and X_2 are fixed while X_3 varies, are parallel, since they are concurrent with $[X_1, X_2, 0]$ and $[0, 0, 1]$. In particular, the line $[1, 0, -x_1]$ is parallel to the axis $[1, 0, 0]$ and $[0, 1, -x_2]$ to $[0, 1, 0]$. These four lines form a parallelogram whose vertices are

$$(0, 0), \quad (x_1, 0), \quad (0, x_2), \quad (x_1, x_2),$$

as in Fig. 12·8A.

From the remark at the end of §11·7, x_1 is the *abscissa* of $(x_1, 0, 1)$, referred to

$$P_0 = (0, 0, 1), \quad P_1 = (1, 0, 1), \quad P_\infty = (1, 0, 0).$$

Comparing this with §8·4, we see that x_1 is actually the *distance* from $(0, 0)$ to $(x_1, 0)$, in terms of the distance to $(1, 0)$ as unit. Similarly x_2 is the distance from $(0, 0)$ to $(0, x_2)$ in terms of the distance to $(0, 1)$ as unit. In affine geometry these two units are, of course, independent; it would be meaningless to regard them as being 'equal'.

The affine theory of conics can be developed by choosing the coordinate axes in convenient positions; e.g. an ellipse for which the coordinate axes are conjugate diameters can be taken in the form

12·81 $$x_1^2 + x_2^2 = 1,$$

while a parabola touching the axis $x_1 = 0$ at $(0, 0)$ is

$$x_2^2 = 2x_1,$$

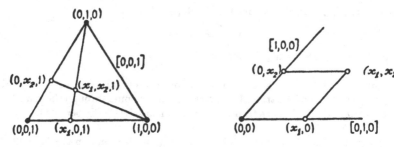

Fig. 12·8A

and a hyperbola whose asymptotes are the coordinate axes is

$$x_1 x_2 = 1.$$

The first of these three equations reminds us of a circle. This is natural enough when we remember that all the *affine* properties of the circle are properties of the ellipse. We may even regard the ellipse as the locus of $(\cos t, \sin t)$, interpreting t as twice the area of the sector from $(1, 0)$ to $(\cos t, \sin t)$. Similarly, the hyperbola

$$x_1^2 - x_2^2 = 1$$

(for which the coordinate axes are a pair of conjugate diameters) is the locus of $(\cosh t, \sinh t)$, where t is twice the area of the sector from $(1, 0)$ to that point (as we may easily verify by integration).

Instead of $[0, 0, 1]$ we might take the line at infinity to be $[1, 1, 1]$. Then $(1, 1, 1)$, the trilinear pole of this line, is the centroid of the triangle of reference, and $(0, 1, 1)$, etc., are the midpoints of the sides. It follows that the ratio of the distances of $(0, x_2, x_3)$ from $(0, 1, 0)$ and $(0, 0, 1)$ (measured in opposite directions) is x_3/x_2 and that the areas of the three triangles joining (x_1, x_2, x_3) to the sides of the triangle of reference are proportional to $x_1 : x_2 : x_3$. Accordingly, these are called *areal* coordinates. [By a well-known result in vector analysis, (x) is the centroid of masses x_1, x_2, x_3 at the vertices of the triangle of reference.]

Since the line at infinity is $x_1 + x_2 + x_3 = 0$, any ordinary point can be normalized so that $x_1 + x_2 + x_3 = 1$. The effect is the same as if we took affine coordinates in three dimensions (analogous to the two-dimensional affine coordinates defined above), restricting attention to the plane whose equation is

$$x_1 + x_2 + x_3 = 1.$$

Then the vectors X_1, X_2, X_3 (p. 180) proceed along the three coordinate axes.

We have seen how the areal coordinates of a point may be measured as areas, but what about the areal coordinates of a line?

12·82 *The areal coordinates of a line are proportional to its distances from the three vertices of the triangle of reference.**

Proof: By similar triangles, these distances have the same ratios in whatever directions they are measured, provided that the direction is the same for all three vertices. In particular, we may measure them along the side $[1, 0, 0]$. Since $[X_1, X_2, X_3]$ meets $[1, 0, 0]$ in the point $(0, X_3, -X_2)$, its distances from $(0, 1, 0)$ and $(0, 0, 1)$ are $-X_2 : X_3$ measured in opposite directions, or $X_2 : X_3$ measured in the same direction. By symmetry, the distances from all three vertices, measured in the same direction, are $X_1 : X_2 : X_3$.

This agrees nicely with the fact that we obtain a parallel line if we increase X_1, X_2, X_3 all by the same amount.

EXERCISES

1. Show that the affine equation

$$a_{11}x_1^2 + 2a_{12}x_1x_2 + a_{22}x_2^2 = 1 \quad (\Delta = a_{12}^2 - a_{11}a_{22})$$

represents a hyperbola if $\Delta > 0$, two parallel lines if $\Delta = 0$; but if $\Delta < 0$ it represents an ellipse or nothing according as a_{11} is positive or negative. In the case of a hyperbola, what is the combined equation for the asymptotes?

2. Find the locus and envelope areal equations of the parabola that touches $[0, 0, 1]$ at $(1, 0, 0)$ and $[1, 0, 0]$ at $(0, 0, 1)$.

3. Show that the areal equation

$$c_1x_1^2 + c_2x_2^2 + c_3x_3^2 = 0$$

represents a parabola if

$$c_1^{-1} + c_2^{-1} + c_3^{-1} = 0.$$

4. Show that $(A_{11} + A_{12} + A_{13}, A_{21} + A_{22} + A_{23}, A_{31} + A_{32} + A_{33})$ is the centre of a non-degenerate conic whose areal equation is $(xx) = 0$ or $[XX] = 0$. Show further that the conic is a hyperbola if $-\Delta$ has the same sign as $A_{11} + A_{22} + A_{33} + 2A_{23} + 2A_{31} + 2A_{12}$, in which case the asymptotes are given by

$$(A_{11} + A_{22} + A_{33} + 2A_{23} + 2A_{31} + 2A_{12})(xx) = (x_1 + x_2 + x_3)^2.$$

12·9 The Euclidean plane: Cartesian and trilinear coordinates. We saw, in §9·1, how a Euclidean metric can be introduced into the affine plane by singling out an elliptic involution on the line at infinity, with the result that a circle has this for its involution of conjugate points on that line. In particular, we derive *rectangular Cartesian coordinates* from affine coordinates by calling 12·81 a circle of radius 1,

---

* Salmon 1879, p. 11.

so that the absolute involution relates $(x_1, x_2, 0)$ and $(y_1, y_2, 0)$, where

12·91
$$x_1 y_1 + x_2 y_2 = 0.$$

The dilatation 12·54 transforms 12·81 into a circle of radius ρ:

$$x_1^2 + x_2^2 = \rho^2.$$

The translation 12·55 (with $x_3 = 1$) transforms this into the general circle

$$(x_1 - c_1)^2 + (x_2 - c_2)^2 = \rho^2,$$

as in classical analytic geometry.

Areal coordinates belong, as we have seen, to the affine plane. They may still be employed after the introduction of a Euclidean metric (when the triangle of reference acquires definite lengths for its sides, say a, b, c); but for many purposes it is desirable to make the transformation

$$x_1 \to a x_1, \quad x_2 \to b x_2, \quad x_3 \to c x_3$$

to *trilinear* coordinates, which are proportional to the distances of the point considered from the three sides, so that the unit point is the incentre (instead of the centroid) and the line at infinity is $[a, b, c]$.

In terms of the angles A, B, C of the triangle of reference ABC, any point on the perpendicular from A to BC is at distances from CA and AB which are proportional to $\cos C : \cos B$. Since

$$a - b \cos C - c \cos B = 0,$$

the point at infinity on this altitude line is $(1, -\cos C, -\cos B)$, and the absolute involution is the degenerate polarity

12·92
$$\begin{cases} x_1 = X_1 - X_2 \cos C - X_3 \cos B, \\ x_2 = -X_1 \cos C + X_2 - X_3 \cos A, \\ x_3 = -X_1 \cos B - X_2 \cos A + X_3, \end{cases}$$

which relates $[1, 0, 0]$ to $(1, -\cos C, -\cos B)$, and so on. Thus the condition for two lines $[X]$ and $[Y]$ to be perpendicular is*

$$X_1 Y_1 + X_2 Y_2 + X_3 Y_3 - (X_2 Y_3 + X_3 Y_2) \cos A$$
$$-(X_3 Y_1 + X_1 Y_3) \cos B - (X_1 Y_2 + X_2 Y_1) \cos C = 0.$$

Now, the left side of this equation is the polarized form of the expression

$$\Omega = X_1^2 + X_2^2 + X_3^2 - 2X_2 X_3 \cos A - 2X_3 X_1 \cos B - 2X_1 X_2 \cos C.$$

---

* Salmon 1854, p. 59.

Since the polarized form of $2\{yX\}\{zX\}$ is $\{yX\}\{zY\} + \{zX\}\{yY\}$, it follows that any two perpendicular lines through either of two points (y) and (z) are conjugate with respect to any conic of the form

$$\{yX\}\{zX\} - \lambda\Omega = 0.$$

Hence any conic with foci (y) and (z) (see §9·7) has such an equation; and by varying λ we obtain a range of *confocal* conics.

Making (y) and (z) coincide, we obtain the equation

12·93 $$\{zX\}^2 - \lambda\Omega = 0 \quad (\lambda > 0)$$

for a circle with centre (z). Different values of λ yield a system of concentric circles. The poles of any fixed line with respect to such a range of circles form a range of points whose parameter λ is a linear function of the distance of such a point from the centre (z). In fact, as this distance vanishes with λ, it is actually proportional to λ. By 9·41, it is also proportional to the square of the radius; but the condition for the circle 12·93 to touch the line $[1, 0, 0]$ is

$$z_1^2 - \lambda = 0.$$

Hence, if z_1, z_2, z_3 are the actual distances of (z) from the three sides of the triangle of reference, the circle with centre (z) and radius ρ is precisely

$$\{zX\}^2 - \rho^2\Omega = 0.$$

In particular, the incircle (for which $z_1 = z_2 = z_3 = \rho = r$) is

$$(X_1 + X_2 + X_3)^2 - \Omega = 0$$

or

$$X_2 X_3 \cos^2 \tfrac{1}{2}A + X_3 X_1 \cos^2 \tfrac{1}{2}B + X_1 X_2 \cos^2 \tfrac{1}{2}C = 0$$

or

$$x_1^{1/2} \cos \tfrac{1}{2}A \pm x_2^{1/2} \cos \tfrac{1}{2}B \pm x_3^{1/2} \cos \tfrac{1}{2}C = 0,$$

and the circumcircle (for which $\rho = R$ and $z_1 = R \cos A$, etc.) is

$$(X_1 \cos A + X_2 \cos B + X_3 \cos C)^2 - \Omega = 0$$

or

$$(X_1 \sin A)^{1/2} \pm (X_2 \sin B)^{1/2} \pm (X_3 \sin C)^{1/2} = 0$$

or

$$(aX_1)^{1/2} \pm (bX_2)^{1/2} \pm (cX_3)^{1/2} = 0$$

or

$$ax_2 x_3 + bx_3 x_1 + cx_1 x_2 = 0.$$

EXERCISES

1. Using Cartesian coordinates, identify the absolute involution 12·91 with the degenerate polarity

$$x_1 = X_1, \quad x_2 = X_2, \quad x_3 = 0.$$

Deduce the condition for lines

$$X_1 x_1 + X_2 x_2 + X_3 = 0 \quad \text{and} \quad Y_1 x_1 + Y_2 x_2 + Y_3 = 0$$

to be perpendicular:

$$X_1 Y_1 + X_2 Y_2 = 0.$$

2. Obtain the Cartesian envelope equation

$$(z_1 X_1 + z_2 X_2 + X_3)^2 - \rho^2(X_1^2 + X_2^2) = 0$$

for the circle with centre (z_1, z_2) and radius ρ.

3. Verify the degeneracy of 12·92.

4. If the trilinear equation $\Sigma\Sigma a_{ij} x_i x_j = 0$ represents a pair of lines, prove that the condition for these lines to be perpendicular is

$$a_{11} + a_{22} + a_{33} - 2a_{23} \cos A - 2a_{31} \cos B - 2a_{12} \cos C = 0.$$

5. Prove that the feet of the perpendiculars from any point on the circumcircle to the sides of the triangle lie on a line (Simson line).

6. Find trilinear coordinates for the midpoints of the sides and the feet of the altitudes. Verify that they satisfy the equation

$$(ax_1 + bx_2 + cx_3)(x_1 \cos A + x_2 \cos B + x_3 \cos C)$$
$$-2(ax_2 x_3 + bx_3 x_1 + cx_1 x_2) = 0$$

or

$$x_1^2 \sin 2A + x_2^2 \sin 2B + x_3^2 \sin 2C$$
$$- (x_1 \sin A + x_2 \sin B + x_3 \sin C)(x_1 \cos A + x_2 \cos B + x_3 \cos C) = 0$$

(the nine-point circle).

7. By examining the involution of conjugate points on the line at infinity $ax_1 + bx_2 + cx_3 = 0$, verify that any circle may be expressed in the form

$$(ax_1 + bx_2 + cx_3)(Y_1 x_1 + Y_2 x_2 + Y_3 x_3) - \lambda(ax_2 x_3 + bx_3 x_1 + cx_1 x_2) = 0.$$

8. Show that the incentre, centroid, circumcentre, orthocentre, and nine-point centre of the triangle of reference ABC have trilinear coordinates

$$(1, 1, 1), \quad (1/a, 1/b, 1/c), \quad (\cos A, \cos B, \cos C),$$

$$(\sec A, \sec B, \sec C), \quad (\cos(B - C), \cos(C - A), \cos(A - B)).$$

Verify that the last four all lie on the Euler line

$$[\sin 2A \sin(B - C), \sin 2B \sin(C - A), \sin 2C \sin(A - B)].$$

12·x Isogonal conjugate points. If the angle $\langle ll' \rangle$ between lines l and l' has the same bisectors as the angle $\langle bc \rangle$ between b and c, the lines l and l' are said to be *isogonal* with respect to b and c*. In other words, l and l' are isogonal if $\langle bl \rangle = \langle l'c \rangle$.

In terms of trilinear coordinates, let the bisectors be the lines $[X]$ and $[Y]$ while l and l' join their intersection to the points (x) and (y). We see from 12·41 that the cross-ratio of these two points and two lines is

$$\frac{\{xX\}\{yY\}}{\{yX\}\{xY\}}$$

and from 9·53 that the isogonal property holds if this cross-ratio is equal to -1, that is, if

12·x1 $$\{xX\}\{yY\} + \{yX\}\{xY\} = 0.$$

Applying 9·59 to the triangle of reference ABC, we see from Fig. 9·5D that the bisectors

$$PS \text{ and } QR, \quad QS \text{ and } RP, \quad RS \text{ and } PQ$$

of the angles A, B, C are

$$x_2 = \pm x_3, \quad x_3 = \pm x_1, \quad x_1 = \pm x_2$$

or

$$[0, 1, \pm 1], \quad [\pm 1, 0, 1], \quad [1, \pm 1, 0].$$

Thus the excentres P, Q, R and the incentre S are

12·x2 $(-1, 1, 1) \quad (1, -1, 1), \quad (1, 1, -1), \quad (1, 1, 1).$

Taking $[X]$ and $[Y]$ to be $[0, 1, -1]$ and $[0, 1, 1]$, so that

$$\{xX\} = x_2 - x_3, \quad \{yY\} = y_2 + y_3, \quad \{yX\} = y_2 - y_3, \quad \{xY\} = x_2 + x_3,$$

we deduce from 12·x1 that the lines joining A to (x) and (y) are isogonal with respect to AC and AB if

$$0 = (x_2 - x_3)(y_2 + y_3) + (y_2 - y_3)(x_2 + x_3) = 2(x_2 y_2 - x_3 y_3),$$

that is, if

$$x_2 y_2 = x_3 y_3.$$

If the two points (x) and (y) lie simultaneously on isogonal pairs of lines through all three vertices of the triangle ABC, they are called *isogonal conjugate points*†. In this case we have

12·x3 $$x_1 y_1 = x_2 y_2 = x_3 y_3$$

---

* Altshiller-Court 1952, p. 267.

† Baker 1943, p. 114; Altshiller-Court 1952, p. 273.

and the point (y) can be identified with the (x^{-1}) of §12·7, namely, the common conjugate point of (x) with respect to all the conics 12·79 which pass through the four points 12·x2. For instance, the isogonal conjugate of the centroid (a^{-1}, b^{-1}, c^{-1}) is the 'symmedian point' (a, b, c).*

Another property of isogonal conjugate points can be obtained by comparison with 12·75, which shows that the conic with foci (y) and (z) touches the lines BC, CA, AB if its equation

$$\{yX\}\{zX\} - \lambda\Omega = 0$$

contains no 'square' terms, that is, if

$$y_1 z_1 = y_2 z_2 = y_3 z_3 = \lambda.$$

Hence:

Any pair of isogonal conjugate points

$$F = (x) \quad and \quad F' = (x^{-1})$$

may be described as the foci of a conic inscribed in the triangle ABC.

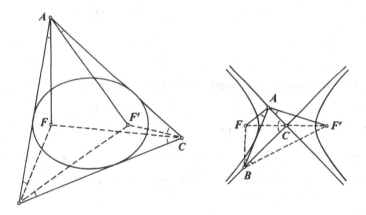

Fig. 12·xA

It follows that, if AB and AC are any two tangents of a conic with foci F and F', the lines AF and AF' (see Fig. 12·xA) are isogonal with respect to those tangents.†

---

* Altshiller-Court 1952, p. 270.
† Robson 1940, p. 286.

EXERCISES

1. Show (in two ways) that the circumcentre and orthocentre of the triangle are isogonal conjugate points.

2. If F varies on the circumcircle, what is the locus of the isogonal conjugate point F'? What kind of conics arise in this case?

3. Prove that every line in the plane of the triangle ABC contains at most one pair of isogonal conjugate points. Which lines (such as $[1, 1, 1]$, or the line at infinity $[a, b, c]$) contain no such pair of points?

The Complex Projective Plane

Many books have been written on a more elaborate geometry where the coordinates of points and lines, instead of being real numbers, are complex numbers.* A large part of the development of Chapter 12 can be carried over. Even the criterion $v/\mu < 0$ for separation (§ 12·4) remains significant, provided we remember that a number cannot be negative without being *real*. (Thus '$v/\mu < 0$' no longer means that μ and v have 'opposite signs'; they might both be imaginary.) The general point collinear with (a) and (b) is $(a + vb)$, where v is any complex number. Three collinear points

$$(a), \quad (b), \quad (a + \mu b)$$

determine a so-called *chain* of points $(a + vb)$ where v/μ is real. Thus the line contains infinitely many chains, and instead of 3·21 we have two points decomposing a chain into two segments. Actually *all but one of the axioms of real projective geometry apply also to complex projective geometry*. The single exception is 3·14, which is false if the four points do not all belong to the same chain. In place of that axiom we have the synthetic definition for a chain, as consisting of three collinear points A, B, C along with the three segments BC/A, CA/B, AB/C. In this terminology, 3·14 states that the chain covers the whole line, which is just what happens in real geometry but not in complex geometry.

Thus everything in Chapter 2 remains valid, but those later results which depend on 3·14 (directly or indirectly) are liable to be contradicted. An interesting example is 3·31, which is contradicted by the following configuration: The nine points defined by the equations

$$x_1^3 + x_2^3 + x_3^3 = x_1 x_2 x_3 = 0$$

---

* For a good exposition of Hamilton's approach to complex numbers, see Robinson 1940, pp. 83–4. It is hardly necessary to add that the words *real* and *imaginary* are picturesque relics of an age when the nature of complex numbers was not yet clearly understood.

viz. $(0, 1, -1),$ $(0, 1, -\omega),$ $(0, 1, -\omega^2),$

$(-1, 0, 1),$ $(-\omega, 0, 1),$ $(-\omega^2, 0, 1),$

$(1, -1, 0),$ $(1, -\omega, 0),$ $(1, -\omega^2, 0),$

where $\omega = e^{2\pi i/3}$, lie by threes on 12 distinct lines in such a way that any two of the points lie on one of the lines. These nine points are the common inflexions of the cubic curves

$$x_1^3 + x_2^3 + x_3^3 + \lambda x_1 x_2 x_3 = 0.$$

The classification of projectivities is actually simpler in complex than in real geometry. For now every projectivity has one or two invariant points, and a line cannot fail to meet a conic. On the other hand, besides the projectivity relating $(a + vb)$ to $(a' + vb')$ there is also an *antiprojectivity* relating $(a + vb)$ to $(a' + \bar{v}b')$, where $\bar{v}$ is the complex conjugates of v. Both these transformations preserve the harmonic relation. Similarly, there is not only a projective collineation 12·51 but also an antiprojective collineation

$$x_i' = \Sigma c_{ij}\bar{x}_j.$$

The projective polarity 12·61 always determines a conic $(xx) = 0$; but there is also an antiprojective polarity

$$X_i = \Sigma a_{ij}\bar{x}_j \quad (a_{ji} = \bar{a}_{ij})$$

whose self-conjugate points form an 'anticonic' $(x\bar{x}) = 0$.

Many results in real geometry are most easily obtained by regarding the real plane as part of the complex plane, so that a real line appears as a chain on a complex line. This method is especially valuable in the theory of circles; for the absolute involution has two invariant points (the *circular points at infinity*, discovered by Poncelet in 1813), and a circle is most simply defined as a conic that passes through these two points.*

---

* Salmon 1879, p. 1.

How to Use Mathematica®

by
George Beck

This *Mathematica* program allows the user to show the scenes of two movie scripts written by H.S.M. Coxeter: *The Arithmetic of Points on a Conic* and *Projectivities*. Accompanying narrations can be found in the files *Arithmetic* and *Projectivities*.

Start up *Mathematica* with some way of reading the narrations. Enter either «Arithmetic.m or «Projectivities.m to load the necessary files.

If you do not have a colour monitor, lines and circles will appear too sketchy. Entering «blacken.m will correct this. If you run out of memory, quit *Mathematica*, start again, and then continue from where you left off.

There are 52 still scenes for *The Arithmetic of Points on a Conic*. To see the nth scene, enter Show[Scene[n]], where n is any one of the numbers from 1 to 52. In the narrations [n] indicates the nth scene.

You can animate the scenes numbered 2, 8, 17, 24, 30, 31, 42, 48, or 49. In the narration these are indicated by [2A], [8A], and so on. To create an animation, enter PreAnimateScene[n, T], where n is one of the numbers just given and T is the number of frames (try 20 at first). The result is a table of graphics; how actually to run the animation will vary from one type of machine to another.

In the script *Projectivities* there are 33 scenes numbered from 61 to 93, any one of which can be animated except for 84 (this is indicated by [84~A] in the narrations). As before, try 20 frames at first, except for the scenes 81, 88, 89 where 4 frames are enough. A representative sampling of the animations is 67, 72, 80, 83, 88, 91.

The file named *Contents* lists all the files (except the scenes) with the functions they contain. Conversely, the file named *Index* lists each function with the file that contains it.

To save memory, functions directly associated with a particular scene are only held in RAM temporarily; so to see how Scene 12 is coded, or to change its default values, use an editor to look at the file Scene12.m.

This program was written using Version 1, but no changes are needed for Version 2. AbsolutePointSize and AbsoluteThickness would keep line colours constant when resizing windows (see globals.m); using Stubs might help to simplify the memory management (see scene.m); and perhaps Block should be changed to Module throughout.

The arithmetic of points on a conic. [1] The example of a thermometer makes it easy to see how the real numbers (positive, zero, and negative) can be represented by the points of a straight line.

[2A]   On the x-axis of ordinary analytic geometry, the number x is represented by the point $(x, 0)$.

[3]   Given any two such numbers, a and b, we can set up geometrical constructions for their sum, ...

[4]   ... difference, ...

[5]   ... product, ...

[6]   ... and quotient.

However, these constructions require a scaffolding of extra points and lines. It is by no means obvious that a different choice of scaffolding would yield the same final results.

The object of the present program is to make use of a circle (or any other conic) instead of the line, so that the constructions can all be performed with a straight edge, and the only arbitrariness is in the choice of the positions of three of the numbers (for instance, 0, 1, and 2).

[7]   Although this is strictly a chapter in projective geometry, let us begin with a prologue in which the scale of the abscissa on the x-axis is transferred to a circle by the familiar process of stereographic projection.

A circle of any radius (say 1, for convenience) rests on the x-axis at the origin 0, and the numbers are transferred from this axis to the circle by lines drawn through the opposite point.

[8A]   That is, the point at the top. In this manner, a definite number is assigned to every point on the circle except the topmost point itself.

[9]   Since the numbers 10, 100, 1000, ... come closer and closer to this point on one side and the numbers -10, -100, -1000, ... come closer and closer on the other side, ...

[10]   ... it is natural to assign the special symbol ∞ to this exceptional point: the only point for which no proper number is available.

The tangent at this exceptional point is, of course, parallel to the x-axis; that is, parallel to the tangent at the point 0.

[11]   Having transferred all the numbers to the circle, we can forget about the x-axis; but the tangent at the point ∞ will play an important role in the construction of sums.

[12]   For instance, there is one point on this tangent which lies on the line joining points 1 and 2, also on the line joining 0 and 3, and on

the line joining -1 and 4. We notice that these pairs of numbers all have the same sum:

$$1 + 2 = 0 + 3 = -1 + 4 = 3.$$

[13]   Similarly, the tangent at 1 meets the tangent at ∞ in a point which lies on the lines joining 0 and 2, -1 and 3, -2 and 4, in accordance with the equations

$$1 + 1 = 0 + 2 = -1 + 3 = -2 + 4.$$

These results could all be verified by elementary analytic geometry, but there is no need to do this, because we shall see later that a general principle is involved.

[14]   Having finished the Euclidean prologue, let us see how far, we can go with the methods of projective geometry (see § 11·2). Let symbols 0, 1, ∞ be assigned to any three distinct points on a given conic.

[15]   There is a certain line through 0 concurrent with the tangents at ∞ and 1; let this line meet the conic again in 2.

(Alternatively, if we had been given 0, 1, 2 instead of 0, 1, ∞, we could have reconstructed ∞ as the point of contact of the remaining tangent from the point where the tangent at 1 meets the line 02.)

We now have the beginning of a geometrical interpretation for all the real numbers.

[16]   To obtain 3, we join 1 and 2, see where this line meets the tangent at ∞, join this point of intersection to 0, and assign the symbol 3 to the point where this line meets the conic again. Thus the line joining 0 and 3, and the line joining 1 and 2, both meet the tangent at ∞ in the same point.

[17A]   More generally, we define addition in such a way that two pairs of points have the same sum if their joins are concurrent with the tangent at the point ∞.

[18]   In other words, we define the sum, $a + b$, of any two points a and b, to be the remaining point of intersection of the conic with the line joining 0 to the point where the tangent at ∞ meets the join of a and b. To justify this definition we must make sure that it agrees with our usual requirements for the addition of numbers: the commutative law

$$a + b = b + a,$$

a unique solution for every equation of the form

$$x + a = c,$$

and the associative law

$$(a + b) + c = a + (b + c).$$

The commutative law is satisfied immediately, as our definition for $a + b$ involves a and b symmetrically.

[19]   The equation $x + a = c$ is solved by choosing x so that x and a have the same sum as 0 and c.

[20]   Thus the only possible cause of trouble is the associative law; we must make sure that, for any three points a, b, c (not necessarily distinct), the sum of $a + b$ and c is the same as the sum of a and $b + c$.

[21]   For this purpose, we make use of a special case of PASCAL'S THEOREM, which says that, if $ABCDEF$ is a hexagon inscribed in a conic, the pairs of opposite sides (namely AB and DE, BC and EF, CD and FA) meet in three points that lie on a line, called the *Pascal line* of the given hexagon.

[22]   In 1639, when Blaise Pascal was sixteen years old, he discovered this theorem as a property of a circle.

[23]   He then deduced the general result by joining the circle to a point outside the plane by a cone and then considering the section of this cone by an arbitrary plane.

We do not know how he proved this property of a hexagon inscribed in a circle, because his original treatise was lost, but we do know how he might have done it, using only the first three books of Euclid's "Elements". In our own time, an easier proof can be found in any textbook on projective geometry.

[24A]   Each hexagon has its own Pascal line. If we fix five of the six vertices and let the sixth vertex run round the conic, we see the Pascal line rotating about a fixed point.

[25]   If this fixed point is outside the conic, we can stop the motion at a stage when the Pascal line is a tangent. This is the special case that concerns us in the geometrical theory of addition.

[26]   The hexagon

$$a \quad b \quad c \quad a+b \quad 0 \quad b+c$$

shows that the sum of $a + b$ and c is equal to the sum of a and $b + c$.

[27]   Beginning with 0, 1, and ∞, we can now construct the remaining positive integers

$$2 = 1 + 1,$$
$$3 = 2 + 1,$$
$$4 = 3 + 1,$$

and so, ...

[28]   ... as well as the negative integers -1, -2, -3, ... given by

$$0 = -1 + 1,$$
$$0 = -2 + 2,$$
$$0 = -3 + 3,$$

and so on, or by ...

[29]
$$1 = -1 + 2,$$
$$1 = -2 + 3,$$
$$1 = -3 + 4,$$

and so on.

[30A]   By fixing b while letting a vary, we obtain a vivid picture of the transformation that adds b to every number a. The points and $a + b$ chase each other round the conic, irrespective of whether b happens to be positive ...

[31A]   ... or negative.

[32]   In our construction for the point 2, we tacitly assumed that the tangent at 1 can be regarded as the join of 1 and 1.

[33]   More generally, the join of a and b meets the tangent at ∞ in a point from which the remaining tangent has, for its point of contact, a point x such that

$$x + x = a + b,$$

namely,

$$(a + b)/2,$$

which is the arithmetic mean (or average) of a and b.

[34]   This result holds not only when $a + b$ is even but also when $a + b$ is odd; for instance, when a and b are consecutive integers. In this way we can interpolate $\frac{1}{2}$ between 0 and 1, $1\frac{1}{2}$ between 1 and 2, and so on.

[35]   We shall find it convenient to work in the scale of 2 (or binary scale), so that the number 2 itself is written as 10, one half as .1, one quarter as .01, three quarters as .11, and so on.

We can now interpolate

[36]   1.1 between 1 and 10, ...
[37]   1.01 between 1 and 1.1, ...
[38]   1.011 between 1.01 and 1.1, ...
[39]   1.01101 between 1.011 and 1.0111,

and so on.

In fact, we can construct a point for every number that can be expressed as a terminating "decimal" in the binary scale. By a limiting process, we can thus theoretically assign a position to every real number.

For instance, the square root of two, being (in the binary scale)

$$1.0110101000001\ldots,$$

is the limit of a certain sequence of constructible numbers:

$$1, 1.01, 1.011, 1.01101, \ldots.$$

[40] Conversely, by a process of repeated bisection, we can assign a binary "decimal" to any given point on the conic. (The "but one" is, of course, the point to which we arbitrarily assigned the symbol ∞.)

[41] We can now define multiplication in terms of the same three points 0, 1, and ∞.

Two pairs of points have the same product if their joins are concurrent with the line joining 0 and ∞.

[42A] The geometrical theory of projectivities (§ 11·5) is somewhat too complicated to describe here, so let us be content to remark that, if we pursued it, we could prove that our definition for addition is consistent with this definition for multiplication.

[43] The product is positive if the point of concurrence is outside, negative if it is inside the conic.

[44] In other words, we define the product, ab, of any two points a and b on the conic, to be the remaining point of intersection of the conic with the line joining 1 to the point where the line joining 0 and ∞ meets the line joining a and b.

Of course, the question arises as to whether this definition agrees with our usual requirements for the multiplication of numbers: the commutative law

$$ab = ba,$$

a unique solution for every equation of the form

$$ax = c$$

(with a not equal to 0), and the associative law

$$(ab)c = a(bc)$$

The commutative law is satisfied immediately, as our definition for ab involves a and b symmetrically.

[45] The equation $ax = c$ is solved by choosing x so that a and x have the same product as 1 and c.

[46] Finally, another application of Pascal's Theorem suffices to show the associative law.

[47] That is, for any three points a, b, c, the product of ab and c is equal to the product of a and bc. In fact, the appropriate hexagon is

$$a \quad b \quad c \quad ab \quad 1 \quad bc.$$

[48A] By fixing b while letting a vary, we obtain a vivid picture of the transformation that multiplies every number by b. If b is positive, the points a and ab chase each other around the conic;

[49A] but if b is negative, they go round in opposite directions.

[50] The familiar identity

$$2 \times 2 = 4$$

is illustrated by the concurrence of the tangent at 2 with the line joining 1 and 4 and the line joining 0 and ∞.

[51] More generally, if a and b are any two numbers having the same sign, the join of the corresponding points meets the line joining 0 to ∞ in a point from which the two tangents have, for their points of contact, points x such that

$$x^2 = ab,$$

namely plus or minus the square root of ab, where the square root of ab is the geometric mean of a and b.

[52] Settng $a = 1$ and $b = 2$, we obtain a construction for the square root of two without having recourse to any limiting process. In fact, we have finite constructions for all the "quadratic" numbers commonly associated with Euclid's straight-edge and compasses.

Projectivities. One of the most fruitful ideas of the nineteenth century is that of one-to-one correspondence. It is well illustrated by the example of cups and saucers. Suppose we have about a hundred cups and about a hundred saucers and wish to know whether the number of cups is actually equal to the number of saucers. This can be determined, without counting, by the simple device of putting each cup on a saucer, that is, by establishing a one-to-one correspondence between the cups and saucers.

In our first application of this idea to plane geometry, the cups are points, the saucers are lines and the relation "cup on saucer" is incidence. As we know, a line is determined by any two of its points and is of unlimited extent. We say that a point and a line are "incident" if the point lies on the line, that is, if the line passes through the point. It is natural to ask whether the number of points on a line is actually equal to the number of lines through a point. In ordinary geometry both numbers are infinite, but this fact need not trouble us—if we can establish a one-to-one correspondence between the points and lines, there are equally many of each.

[61] The set of all points on a line o is called a *range* and the set of all lines through a point O is called a *pencil*. If the line o and the point O are not incident, we can establish an *elementary correspondence* between the range and the pencil by means of the relation of incidence. Each point X of the range lies on a corresponding line x of the pencil. The range is a section of the pencil (namely, the section by the line o) and the pencil projects the range (from the point O).

In our picture, the range is represented by a red point X moving along a fixed line o (which, for convenience, is taken to be horizontal) and the pencil is represented by a green line x rotating around a fixed point O.

There is evidently a green line for each position of the red point, but we must admit that for some positions of the green line the red point cannot be seen because it is too far away; in fact, when the green line is parallel to o (that is, horizontal), the red point is one of the ideal "points at infinity" which we agree to add to the ordinary plane so as to make the projective plane. Without this ideal point, our elementary correspondence would not be one-to-one—the number of points in the range would be one less than the number of lines in the pencil. In other words, the postulation of ideal points makes it possible for us to express the axioms for the projective plane in such a way that they remain valid when we consistently interchange the words "point" and "line" (and consequently also certain other pairs of words such as "join" and "meet", "on" and "through", "collinear" and "concurrent", and so forth). It follows that the same kind of interchange can be made in all the theorems that can be deduced from the axioms. This principle of *duality* is characteristic of projective geometry. In the plane we interchange points and lines. In space, the same principle enables us to interchange points and planes, while lines remain lines.

When we regard the elementary correspondence as taking us from the point X to the line x, we write the capital X before the small x. The *inverse* correspondence, from x to X, is denoted by the same sign with the small x before the capital X. If $A, B, C, \ldots$ are particular positions of X, and $a, b, c, \ldots$ of x, we write all these letters before and after the sign, taking care to keep them in their corresponding order (which need not be the order in which they appear to occur in the figure).

This notation enables us to exhibit the principle of duality as the possibility of consistently interchanging capital and small letters.

By combining two elementary correspondences, one relating a range to a pencil, and the other a pencil to a range, we obtain a *perspectivity*. This either relates two ranges that are different sections of one pencil, or two pencils which project one range from different centres.

[62, 63]   In the former case, two of the symbols with one bar can be abbreviated to one with two bars, or, if we wish to specify the point O that carries the pencil, we put O above the two bars.

[64, 65]   In the latter case (when two pencils project one range from different centres), the two symbols with one bar are again abbreviated to one with two bars, and if we wish to specify the line o that carries the range, we put o above the bars.

We can easily go on to combine three or more elementary correspondences, but then we prefer not to increase the complication of the symbols. Instead, we retain the simple symbol (with just one bar) for the product of any number of elementary correspondences. Such a transformation is called a *projectivity*. Thus elementary correspondences and perspectivities are the two simplest instances of a projectivity.

[66, 67]   The product of three elementary correspondences is the

simplest instance of a correspondence relating a range to a pencil in such a way that the range is not merely a section of the pencil.

[68] The product of four elementary correspondences, being the product of two perspectivities, shares with a simple perspectivity the property of relating a range to a range or a pencil to a pencil. Now there is the interesting possibility that the initial and final range (or pencil) may be on the same line (or through the same point). We see two moving red points X and Z, on o, related by perspectivities from O and M to an auxiliary red point Y on m. When X reaches C, on m, we have another invariant point; the three red points all come together. Such a projectivity, having two distinct invariant points, is said to be *hyperbolic*.

[69] On the other hand, the three lines o, m, and OM may all meet in a single point C, so that F coincides with C and there is only one invariant point. Such a projectivity is said to be *parabolic*.

[70] A third possibility is an *elliptic* projectivity which has no invariant point; but this is more complicated, requiring three perspectivities (i.e., six elementary correspondences). The centers of the three perspectivities are S, E, and Q. The green lines, rotating around these points, yield four red points. Two of the red points chase each other along the bottom line AD. These two points are related by the elliptic projectivity. This, however, is not the most general elliptic projectivity. There is a special feature arising from the fact that the points S, E, Q lie on the sides of the green triangle. When one of the two red points is at A, the other is at D, and vice versa: the projectivity *interchanges* A and D and is consequently called an *involution*. Thus we are watching an *elliptic involution*.

[71] Looking closely, we see that it not only interchanges A and D but also interchanges every pair of related points. For instance, it interchanges E with B (on SQ). An important theorem tells us that, for any four collinear points A, B, D, E, there is just one involution which interchanges A with D and B with E. We denote it by $(AD)(BE)$. At any instant, the two red points are a pair belonging to this involution. Call them C and F. We now have three pairs of points, AD, BE, CF, on the bottom dark blue line, all belonging to one involution. The other lines form the six sides of a *complete quadrangle PQRS*, which consists of four points (no three collinear) and the six lines that join them in pairs. Two sides are said to be *opposite* if their point of intersection is not a vertex; for instance, SP and QR are a pair of opposite sides.

[72] We see now that the six points named on the bottom dark blue line are sections of the six sides of the quadrangle and that each related pair comes from a pair of opposite sides. Accordingly the six points, paired in this particular way, are said to form a *quadrangular set*. Here is another version of the quadrangle $PQRS$ and the corresponding quadrangular set AD, BE, CF. As before, CF is a pair of the involution $(AD)(BE)$.

[73] This remains true when we move the bottom dark blue line to a new position so that D coincides with A and E with B. Now A and B are invariant points, and we have a *hyperbolic* involution $(AA)(BB)$, which still interchanges C and F.

The quadrangular set of six points has become a *harmonic set* of four points. We say that C and F are harmonic *conjugates* of each other with respect to A and B and that the four points satisfy the relation $H(AB, CF)$.

This means that there is a quadrangle $PQRS$ having two opposite sides through A, two opposite sides through B, while one of the remaining two sides passes through C and the other through F.

[74] Given A, B, and C, we can construct F by drawing a triangle SRQ whose sides pass through these three points respectively.

[75] Let AS meet BR in P; then PQ meets AB in F. Of course, the hyperbolic involution $(AA)(BB)$ can still be constructed as the product of three perspectivities (with centers S, B, Q).

[76] But the invariant points A and B enable us to replace these three perspectivities by two, with centers O (where AP meets CQ) and P.

[77] Another product of two perspectivities relates ranges on two distinct lines. The Fundamental Theorem of Projective Geometry (§ 4·2) tells us that a projectivity relating ranges on two such lines is uniquely determined by any three points of the first range and the corresponding three points of the second. There are, of course, many ways to construct the projectivity as the product of two or more perspectivities, but the final result will always be the same.

For instance, there is a unique projectivity relating AED on the first line to BDC on the second. This means that, for any point X on DE there is a definite point Y on CD.

[78] The simplest way to construct this projectivity is by means of perspectivities from B and A, so that X is first related to Z on EC and then to Y on CD. We can regard XYZ as a variable triangle whose vertices run along fixed lines DE, EC, CD, while the two sides YZ and ZX rotate around fixed points A and B. The third side joins the projectively related points X and Y.

[79] This construction remains valid when A and B are of general position, instead of lying on the lines that carry the related ranges. Let AB meet DE in I, and CD in J. Now we have a construction for the unique projectivity that relates IED to JDC. As before, the vertices of the variable triangle XYZ run along fixed lines DE, EC, CD while the two sides YZ and ZX rotate around the fixed points A and B. The possible positions for the third side XY include, in turn, each of the five sides of the pentagon $ABCDE$.

[80] Carefully watching this line XY, we see that it envelops a beautiful curve.

This is the same kind of curve that was constructed quite differently by Menaechmus about 340 B.C. Since that time it has been known everywhere as a *conic*. One important property is that a conic is uniquely determined by any five of its tangents and that these may be any five lines of which no three are concurrent.

[81] Since the possible positions for our variable line XY include, in turn, each side of the pentagon $ABCDE$, we call its envelope the conic *inscribed* in this pentagon.

[82] To sum up: Let Z be a variable point on the diagonal CE of a given pentagon $ABCDE$. Then the point X, where ZB meets DE, and the point Y, where ZA meets CD, determine a line XY whose envelope is the inscribed conic.

[83] For any particular position of Z (on CE), we see a hexagon $ABCYXE$ whose six sides all touch the conic. The three lines AY, BX, CE, which join pairs of opposite vertices, are naturally called *diagonals* of the hexagon. Thus, if the diagonals of a hexagon are concurrent, the six sides all touch a conic. Conversely, if all the sides of a hexagon touch a conic, five of them can be identified with the lines DE, EA, AB, BC, CD. Since the given conic is the only one that touches these fixed lines, the sixth side must coincide with one of the lines XY that we have constructed. We thus have BRIANCHON'S THEOREM: *If a hexagon is circumscribed about a conic, the three diagonals are concurrent.*

[84~A] All these results can, of course, be dualized. (Now all the letters that we use are lower-case, representing lines.)

[85] For any pentagon $abcde$, whose vertex $a \cdot b$ is joined to $d \cdot e$ by i, and to $c \cdot d$ by j; there is a unique projectivity relating ied to jdc.

[86] The sides of the variable triangle xyz rotate about fixed points $d \cdot e$, $e \cdot c$, $c \cdot d$ while the two vertices $y \cdot z$ and $z \cdot x$ run along the fixed lines a and b. The possible positions for the third vertex $x \cdot y$ include, in turn, each of the five vertices of the pentagon.

[87] Carefully watching this moving point $x \cdot y$, we see that it traces out a curve through these five fixed points (no three concurrent). What is this curve, the dual of a conic?

[88] One of the many possible definitions for a conic exhibits it as a self-dual figure, with the interesting result that the dual of a conic (regarded as the envelope of its tangents) is again a conic (regarded as the locus of the points of contact of these tangents).

[89] Thus the locus of the point $x \cdot y$ is a conic, and this is the only conic that can be drawn through the five vertices of the pentagon.

[90] To sum up: Let z be a variable line through the intersection $c \cdot e$ of two non-adjacent sides of a given pentagon $abcde$. Then the line x, which joins $z \cdot b$ to $d \cdot e$, and the line y, which joins $z \cdot a$ to $c \cdot d$, determine a point $x \cdot y$ whose locus is the circumscribed conic.

[91] The hexagon $abcyxe$ which, for convenience, we rename $abcdef$, yields the dual of Brianchon's Theorem, namely PASCAL'S THEOREM:

If abcdef is a hexagon inscribed in a conic, the points a·d, b·d, c·f (where pairs of opposite sides meet) are collinear.

The hexagon that we see is, perhaps, unusual, because its sides cross one another. From the standpoint of projective geometry, this feature is irrelevant. A convex hexagon *abcdef* would serve just as well, but the "diagonal points" would be inconveniently far away. Another natural observation is that our conic looks like the familiar circle. In fact, this famous theorem was first proved for a circle in 1639, when its discoverer, Blaise Pascal, was only sixteen years old. Nobody knows just how he did it, because his original treatise has been lost.

[92] There is no possible doubt about how he deduced the analogous property of the general conic. He joined the circle and lines to a point outside the plane, obtaining a cone and planes. Then he took the section of this solid figure by an arbitary plane.

[93] In this way the conic appears in one of its most ancient aspects: as the section of a circular cone by a plane of general position.

NOTE: All files on the DOS disk are located in the COXETER subdirectory.

Bibliography

Nathan Altshiller-Court, 1952. *College geometry*, Barnes and Noble, New York.

Apollonius of Perga, 1891. *Conicorum*, Heiberg, Leipzig.

Emil Artin, 1940. *Coordinates in affine geometry*, Reports of a Mathematical Colloquium (Notre Dame), vol. **2.2**.

H.F. Baker, 1929. *Principles of geometry, vol.* I, Cambridge Univ. Press.

H.F. Baker, 1930. *Principles of geometry, vol.* II, Cambridge Univ. Press.

H.F. Baker, 1943. *Introduction to plane geometry*, Cambridge Univ. Press.

Peter Borwein and W.O.J. Moser, 1990. *A survey of Sylvester's problem and its generalizations*, Aequationes Math. **40**, 111–135.

Michel Chasles, 1865. *Traité des sections coniques*, Gauthier-Villars, Paris.

J.L. Coolidge, 1945. *A history of the conic sections and quadric surfaces*, Clarendon, Oxford.

H.S.M. Coxeter, 1965. *Non-Euclidean geometry* (5th ed.), Univ. of Toronto Press.

H.S.M. Coxeter, 1969. *Introduction to geometry* (2nd ed.), Wiley, New York.

H.S.M. Coxeter, 1973. *Regular polytopes* (3rd ed.), Dover, New York.

H.S.M. Coxeter, 1991. *Regular complex polytopes* (2nd ed.), Cambridge Univ. Press.

Luigi Cremona, 1960. *Elements of projective geometry*, Dover, New York.

J.D.H. Donnay, 1945. *Spherical trigonometry after the Cesàro method*, Interscience, New York.

C.V. Durell, 1931. *Projective geometry*, Macmillan, London.

Federigo Enriques, 1930. *Leçons de géométrie projective*, Gauthier-Villars, Paris.

H.G. Forder, 1927. *Foundations of Euclidean and Non-Euclidean geometry*, Cambridge Univ. Press.

H.G. Forder, 1931. *Higher course geometry*, Cambridge Univ. Press.

H.G. Forder, 1947. *The cross and the foundations of Euclidean geometry*, Math. Gazette **31**, 227–233.

W.C. Graustein, 1930. *Introduction to higher geometry*, Macmillan, New York.

H.W. Guggenheimer, 1967. *Plane geometry and its groups*, Holden-Day, San Francisco.

G.H. Hardy, 1925. *A course of pure mathematics* (4th ed.), Cambridge Univ. Press.

Lothar Heffter and C. Koehler, 1905. *Lehrbuch der analytischen Geometrie, vol.* I, Teubner, Leipzig.

L.O. Hesse, 1897. *Lehrbuch über analytische Geometrie des Raumes*, Teubner, Leipzig.

Gerhard Hessenberg, 1930. *Grundlagen der Geometrie*, de Gruyter, Berlin.

David Hilbert, 1930. *Grundlagen der Geometrie*, Teubner, Leipzig.

W.V.D. Hodge and Daniel Pedoe, 1947. *Methods of algebraic geometry, vol.* I, Cambridge Univ. Press.

T.F. Holgate, 1930. *Projective pure geometry*, Macmillan, New York.

R.A. Johnson, 1929. *Modern geometry*, Houghton Mifflin, Cambridge, MA.

D.N. Lehmer, 1917. *An elementary course in synthetic projective geometry*, Boston, MA.

F.W. Levi, 1929. *Geometrische Konfigurationen*, Hirzel, Leipzig.

Sophus Lie, 1893. *Vorlesungen über continuierliche Gruppen*, Teubner, Leipzig.

F.S. Macaulay, 1906. *Geometrical conics* (2nd ed.), Cambridge Univ. Press.

G.B. Mathews, 1914. *Projective geometry*, Longmans Green, London.

N.S. Mendelsohn, 1944. *Multiplication by addition and reciprocation*, Amer. Math. Monthly **51**, 171.

A.F. Möbius, 1827, *Der barycentrische Calcul*, Barth, Leipzig.

E.H. Neville, 1960. *The nine-point conic of a convex quadrangle*, Math. Gazette **44**, 214–215.

C.W. O'Hara and D.R. Ward, 1937. *An introduction to projective geometry*, Clarendon, Oxford.

Blaise Pascal, 1908. (*Euvres, vol.* I (Brunschvicg and P. Boutroux, eds.), Hachette, Paris.

Moritz Pasch and Max Dehn, 1926. *Vorlesungen über neuere Geometrie*, Springer, Berlin.

D.K. Picken, 1925. *Euclidean geometry of angle*, Proc. London Math. Soc. (2) **23**, 43–55.

Mario Pieri, 1899. *I principii della geometria de posizione, composti in sistema logico deduttivo*, Mem. Reale Accad. Sci. Torino (2) **48**, 1–62.

J.V. Poncelet, 1865. *Traité des propriétés projectives des figures* (2nd ed.), Gauthier-Villars, Paris.

Theodor Reye, 1923. *Die Geometrie der Lage* (6th ed.), Rumpler, Hanover.

G. deB. Robinson, 1940. *The foundations of geometry*, Univ. of Toronto Press.

Alan Robson, 1940. *An introduction to analytical geometry, vol.* I, Cambridge Univ. Press.

Alan Robson, 1947. *An introduction to analytical geometry, vol.* II, Cambridge Univ. Press.

Bertrand Russell, 1930. *Introduction to mathematical philosophy* , (2nd ed.), Alen Unwin, London.

Bertrand Russell, 1937. *Principles of mathematics* , (2nd ed.), Allen and Unwin, London; W.W. Norton, New York.

George Salmon, 1879. *A treatise on the higher plane curves* (3rd ed.), Hodges, Foster, and Figgis, Dublin.

George Salmon, 1954. *A treatise on conic sections* (6th ed.), Chelsea, New York.

Charles Smith, 1921. *An elementary treatise on conic sections*, Macmillan, London.

K.G.C. von Staudt, 1847. *Geometrie der Lage*, F. Korn, Nürnberg.

K.G.C. von Staudt, 1857. *Beiträge zur Geometrie der Lage*, F. Korn, Nürnberg.

Oswald Veblen and J.W. Young, 1910. *Projective geometry, vol.* I, Blaisdell, New York.

Oswald Veblen and J.W. Young, 1918. *Projective geometry, vol.* II, Blaisdell, New York.

R.J. Walker, 1946. *Conics*, Amer. Math. Monthly **53**, 538–539.

A.N. Whitehead, 1906. *Axioms of projective geometry*, Cambridge Univ. Press.

Oswald Wyler, 1952. *Order and topology in projective planes*, American Journal of Mathematics, **74**, 656–666.

J.W. Young, 1930. *Projective geometry* (4th Carus Monograph), Open Court, Chicago.

Index

Printed in the United States
By Bookmasters